New Perspectives in Human Embryology

Edited by Bin Wu

Published in London, United Kingdom

New Perspectives in Human Embryology
http://dx.doi.org/10.5772/intechopen.1001771
Edited by Bin Wu

Contributors
Bin Wu, Christopher Edwards, Guanglei Li, Hossein Azizi, Qing Zhou, Suzhen Lu, Timothy J. Gelety, Xingxu Huang, Xue Feng, Yutong Wang, Yu Zhang, Zahra Hasani Mahforoozmahalleh

First published in London, United Kingdom, 2024 by IntechOpen
IntechOpen is the global imprint of INTECHOPEN LIMITED, registered in England and Wales, registration number: 11086078, 167-169 Great Portland Street, London, W1W 5PF, United Kingdom

British Library Cataloguing-in-Publication Data
A catalogue record for this book is available from the British Library

Additional hard and PDF copies can be obtained from orders@intechopen.com

New Perspectives in Human Embryology
Edited by Bin Wu
p. cm.

This title is part of the Obstetrics and Gynecology Book Series, Volume 4
Topic: Reproductive and Gynecologic Health
Series Editor: Zouhair O. Amarin
Topic Editor: Courtney Marsh

Print ISBN 978-0-85466-215-9
Online ISBN 978-0-85466-214-2
eBook (PDF) ISBN 978-0-85466-216-6
ISSN 3049-706X

For EU product safety concerns:
IN TECH d.o.o., Prolaz Marije Krucifikse Kozulić 3, 51000 Rijeka, Croatia,
info@intechopen.com or visit our website at intechopen.com.

IntechOpen Book Series

Obstetrics and Gynecology

Volume 4

Aims and Scope of the Series

Obstetrics and Gynecology field typically covers several subspecialties that include maternal-fetal medicine, gynecologic oncology, reproductive endocrinology and subfertility, urogynecology and female pelvic reconstructive surgery, critical care medicine, complex family planning, pediatric and adolescent gynecology, menopausal and geriatric gynecology, and minimally invasive gynecologic surgery.
Apart from being diverse, obstetrics and gynecology is a challenging and demanding specialty. It combines medical and surgical skills.
The sequence of books in this series will have certain characteristics in common that will be recognized as part of the same informative multi-volume book programme.

Meet the Series Editor

Zouhair Amarin is a Professor of Obstetrics and Gynaecology at the Jordan University of Science and Technology. He was previously a lecturer at the University of Glasgow, Scotland, a senior lecturer at the University of Nottingham, England, and the dean of the Faculty of Medicine at Mutah University, Jordan. Professor Amarin is a fellow of the Royal College of Obstetricians and Gynaecologists, and the Faculty of Public Health, London. He holds master's degrees in medical science and medical education. He is a pioneer in IVF and was the first in the world to develop microsurgical epididymis sperm aspiration for clinical use. He also discovered a surgical procedure for critical ovarian hyperstimulation syndrome. Professor Amarin has edited books, authored book chapters, and published more than 130 papers. He is the recipient of eight awards.

Meet the Volume Editor

Bin Wu, Ph.D., HCLD (ABB) is currently a scientific laboratory director at Arizona Center for Reproductive Endocrinology and Infertility, USA. He received his training in genetics and reproductive biology at the NWAU in China and at Cornell University, as well as post-doctorate training at the University of Guelph, Canada. As an embryologist, he joined the Center for Human Reproduction in Chicago. He has memberships in the American Society for Reproductive Medicine and the American Association of Bioanalysts. Also, he has obtained some significant research awards from professional associations. The American College of Embryology recognized Dr. Wu's excellence in the practice of reproductive embryology in July 2014. Dr. Wu has published 50 papers and edited five published books in the field of embryology.

Contents

Preface

Human embryology is the study of early embryo development during the first eight weeks after fertilization and extends to the normal period of gestation (pregnancy), about nine months or 40 weeks. The study of human embryology has a very long history. Early embryology study owes its initial development to the key embryo collections that began in the nineteenth century. The first large collection was that of Carnegie, which allowed researchers to establish a defined set of Carnegie stages based on embryo morphological features. Currently, assistant reproductive technology, especially the development of the in vitro fertilization (IVF) technique, has given us more understanding of the theory and practice of the embryo's formation process, its morphology, and related mechanisms. In the past decade, many innovative technologies and theories have been advanced in human embryology, in particular in human clinical practice. Thus, this book collects five perspective papers in the field of human embryology so that we have the newest knowledge in this area.

The first chapter identifies some novel transcription factors involved in spermatogonial stem cells (SSCs). These new discoveries help us to understand spermatogenesis and to predict the molecular connections and functional characteristics of proteins during SSC activation to form sperm. This is a very interesting paper for human embryology knowledge. Next, three perspective papers in the field of human-assisted reproductive technology describe the application of abnormally fertilized eggs and the associated clinical outcomes, noninvasive testing of preimplantation embryos, and the research progress of correcting pathogenic mutations by base editing. These perspectives will be helpful for human IVF physicians and embryologists to further improve human IVF outcomes. Finally, a perspective chapter discusses a new view of the endocrinology of pregnancy and parturition, describing the regulation mechanism of human parturition through lessons from the literature. Thus, this book provides perspective information for readers, especially embryologists and physicians as well as obstetricians/gynecologists, for their clinical practice.

Great thanks go to all authors who gladly contributed their time and expertise to prepare these outstanding chapters included in this book.

Bin Wu, Ph.D., HCLD (ABB)
Arizona Center for Reproductive Endocrinology and Infertility,
Tucson, Arizona, United States of America

Section 1

Spermatogenesis

Chapter 1

An Overview of Novel Transcription Factors Involved in Spermatogonial Stem Cells

Zahra Hasani Mahforoozmahalleh and Hossein Azizi

Abstract

A unique subset of spermatogonial stem cells (SSCs) initiates and maintains spermatogenesis. These SSCs have unique morphological traits attached to the seminiferous tubules basement membrane. They provide the groundwork for a healthy stem cell system in the testis, which is essential for spermatogenesis and other reproductive functions. The fascinating proteins known as transcription factors (TFs) have a great deal of control over gene expression in all living things. Some TFs are essential to the coordination of the complex dance known as spermatogenesis. Certain mutations in TFs may lead to the disorder of spermatogenesis. Distinguishing these TFs will be helpful to understand spermatogenesis and to locate possible therapeutic targets. In this chapter, we will review the recently identified TFs including E4F1, FoxP4, A-MYB, TCFL5, and TCF3 that play a role in SSCs. Enrich Shiny gene ontology and Cytoscape tools were used to predict the molecular connections and functional characteristics of proteins in order to find the key pathways. Our bioinformatic analysis will help us to understand these new and important connections between the TFs and the remaining gene expression in the protein network.

Keywords: spermatogonial stem cells, transcription factors, spermatogenesis, gene ontology, bioinformatic

1. Introduction

In the male reproductive system, spermatogenesis is one of the most important processes. Spermatogonia stem cells (SSCs) play the most fundamental role in this system [1]. The process of spermatogenesis involves multiple steps involving SSCs that simultaneously undergo self-renewal and differentiation and then undergo sequential cell-fate transitions between mitosis and meiosis to produce highly specialized sperms [2]. The process of spermatogenesis also involves spermiogenesis (a spermatogenesis is the metamorphosis of spermatogonia into spermatozoa and includes all functional and structural changes that occur throughout this process) [3, 4]. During the course of spermatogenesis, cells undergo a number of stages where they progress from spermatogonia to mature sperm [5]. In spermatogenesis, genetic information is passed on to the next generation [6]. When it comes to spermatogenesis, it is very important that proteins are expressed in the proper order and location at crucial stages of the process [7]. Biological evolution is based on the variable expression of genetic information, and an organism's ability to adapt to changes in its environment is largely dependent on the control of gene expression. Gene expression levels, location, and timing are all strictly regulated. Gene regulation is a multifaceted and intricate process that primarily operates at the transcriptional level. Transcription factors bind to certain DNA sequences to modulate chromatin shape and target gene transcription through intricate networks of interacting factors [8]. SSCs are thought to be a promising substitute for injured or compromised spermatogenesis in order to regenerate it [9].

While a number of spermatogonial stem cell regulators have been well studied in rodents, the regulatory mechanisms governing the self-renewal and differentiation of human SSCs remain incompletely understood [10]. The transcription factors Promyelocytic leukemia zinc finger protein (Plzf) and TATA-box binding protein associated factor 4b (Taf4b) have previously been linked to the regulation of SSC functions; these molecules are a part of a complex gene network that determines the fate of SSCs [9]. There are a number of important gene expression regulators in spermatogenesis that have been found, especially in spermiogenesis. For instance, alternate splicing of the cAMP responsive element modulator (CREM) transcripts in pachytene SCs (pacSCs) and delayed translation in round spermatids (rSTs) result in the expression of CREM-τ, a transcription activator that is exclusively produced in rSTs. Spermatogenesis is stopped in mice with CREM gene knockouts (KO) at the early rST stage. Numerous genes involved in a variety of biological activities have their expression directly or indirectly regulated by CREM-τ. These characteristics have led to the recognition of CREM-τ as a crucial regulator of spermiogenesis. Transcription factors (TFs) like TRF2, TAF7L, and RFX2 and RNA-binding proteins (RBPs) like TPAP, MIWI, RNF17, BOULE/BOLL, and GRTH/DDX25 are examples of similar regulators [11]. Whereas a number of spermatogonial stem cell regulators have been well studied in rodents, the regulatory mechanisms governing the self-renewal and differentiation of human SSCs remain incompletely understood [6, 7, 10, 12–17].

We will go over the newly discovered TFs involved in SSCs in this book chapter. Furthermore, the bioinformatic analyses performed resulted in the comprehension and identification of novel and significant associations between these transcription factors and the remaining genes within the protein network.

2. Some novel transcription factors involved in spermatogonial stem cell proliferation and differentiation

2.1 FOXP4

The "Forkhead" motif-bearing Forkhead box (Fox) family of transcription factors is an essential component of human health and sickness that has been preserved throughout evolution. Apoptosis, cell differentiation, aging, embryonic development, glucose and lipid metabolism, and immunological control are all impacted by genes belonging to the Fox family [18]. It has been recently established that the FOXP subfamily is a part of the Fox gene family. In addition to having a forkhead domain, members of the FOXP subfamily also have a zinc finger domain and a leucine zipper motif [19]. Tucked down on human chromosome 6p21.1, forkhead box P4 (FOXP4) is a member of the FOXP subfamily that codes for a 680 amino acid protein [20].

According to recent research, FOXP4 may be involved in the oncogenesis of several tissues, including hepatocellular carcinoma, prostate cancer, and non-small-cell lung cancer. In addition to being well-known for controlling tissue growth or cell differentiation throughout life, the FOXP family (FOXP1-4) is also important for oncogenesis, cell-cycle regulation, and embryonic development. Among them, it has been documented that FOXP4 controls the differentiation and regeneration of lung secretory epithelial cells [21].

Luo et al. examined human testis single-cell sequencing data and discovered that the expression of forkhead box P4 (FOXP4) progressively increased as SSCs developed [10]. Subsequent examination of FOXP4's expression patterns in human testicular tissues showed that it identifies spermatogonia with stem cell potential in a distinct subgroup. In human SSC lines, conditional FOXP4 silencing greatly increased apoptosis and inhibited SSC growth. The expression of FOXP4 was significantly reduced in areas where spermatogenesis was dysregulated. These results suggest that FOXP4 plays a role in human SSC proliferation, which will clarify the processes governing the decisions made about the fate of human SSCs [10].

2.1.1 Analyzing FOXP4 in gene network and bioinformatic tools

The PPI network was visualized using the Cytoscape software version 3.10.1 (https://cytoscape.org/) as shown in **Figure 1A**. BCL6, CTBP1, DHDDS, EZH2, FOXP1, FOXP1-2, FOXP2, FOXP3, GATAD2B, HAND1, HAND2, LCOR, NFATC2, PRR20D, PRR20E, SATB1, SRPX2, TBR1, TINCR, and ZNF609 are the associated genes.

Terms related to the biological process (BP) and molecular function (MF) were revealed using functional enrichment analysis (https://maayanlab.cloud/Enrichr/). Moreover, the biological pathway mediated by related TRANSCRIPTION FACTOR

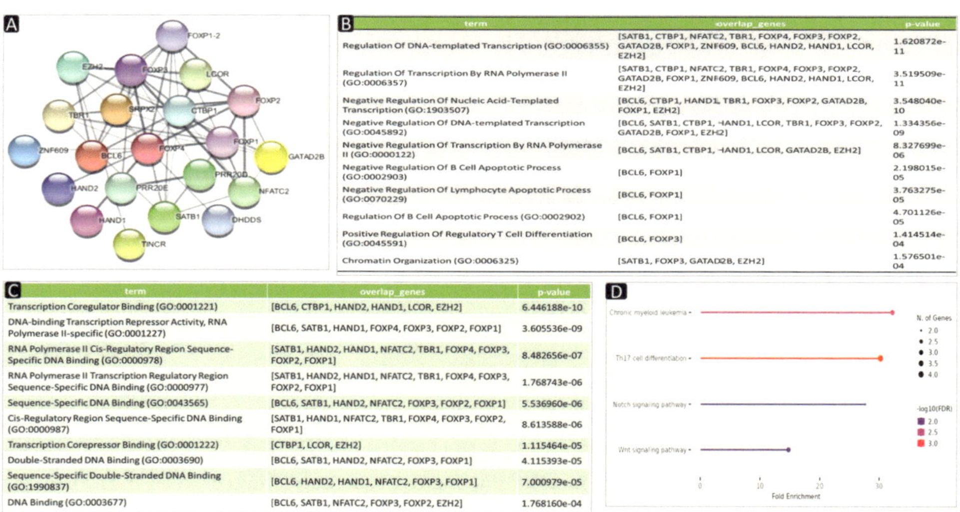

B

term	overlap_genes	p-value
Regulation Of DNA-templated Transcription (GO:0006355)	[SATB1, CTBP1, NFATC2, TBR1, FOXP4, FOXP3, FOXP2, GATAD2B, FOXP1, ZNF609, BCL6, HAND2, HAND1, LCOR, EZH2]	1.620872e-11
Regulation Of Transcription By RNA Polymerase II (GO:0006357)	[SATB1, CTBP1, NFATC2, TBR1, FOXP4, FOXP3, FOXP2, GATAD2B, FOXP1, ZNF609, BCL6, HAND2, HAND1, LCOR, EZH2]	3.519509e-11
Negative Regulation Of Nucleic Acid-Templated Transcription (GO:1903507)	[BCL6, CTBP1, HAND1, TBR1, FOXP3, FOXP2, GATAD2B, FOXP1, EZH2]	3.548040e-10
Negative Regulation Of DNA-templated Transcription (GO:0045892)	[BCL6, SATB1, CTBP1, HAND1, LCOR, TBR1, FOXP3, FOXP2, GATAD2B, FOXP1, EZH2]	1.334356e-09
Negative Regulation Of Transcription By RNA Polymerase II (GO:0000122)	[BCL6, SATB1, CTBP1, HAND1, LCOR, GATAD2B, EZH2]	8.327699e-06
Negative Regulation Of B Cell Apoptotic Process (GO:0002903)	[BCL6, FOXP1]	2.198015e-05
Negative Regulation Of Lymphocyte Apoptotic Process (GO:0070229)	[BCL6, FOXP1]	3.763275e-05
Regulation Of B Cell Apoptotic Process (GO:0002902)	[BCL6, FOXP1]	4.701126e-05
Positive Regulation Of Regulatory T Cell Differentiation (GO:0045591)	[BCL6, FOXP3]	1.414514e-04
Chromatin Organization (GO:0006325)	[SATB1, FOXP3, GATAD2B, EZH2]	1.576501e-04

C

term	overlap_genes	p-value
Transcription Coregulator Binding (GO:0001221)	[BCL6, CTBP1, HAND2, HAND1, LCOR, EZH2]	6.446188e-10
DNA-binding Transcription Repressor Activity, RNA Polymerase II-specific (GO:0001227)	[BCL6, SATB1, HAND1, FOXP4, FOXP3, FOXP2, FOXP1]	3.605536e-09
RNA Polymerase II Cis-Regulatory Region Sequence-Specific DNA Binding (GO:0000978)	[SATB1, HAND2, HAND1, NFATC2, TBR1, FOXP4, FOXP3, FOXP2, FOXP1]	8.482656e-07
RNA Polymerase II Transcription Regulatory Region Sequence-Specific DNA Binding (GO:0000977)	[SATB1, HAND2, HAND1, NFATC2, TBR1, FOXP4, FOXP3, FOXP2, FOXP1]	1.768743e-06
Sequence-Specific DNA Binding (GO:0043565)	[BCL6, SATB1, HAND2, NFATC2, FOXP3, FOXP2, FOXP1]	5.536960e-06
Cis-Regulatory Region Sequence-Specific DNA Binding (GO:0000987)	[SATB1, HAND1, NFATC2, TBR1, FOXP4, FOXP3, FOXP2, FOXP1]	8.613588e-06
Transcription Corepressor Binding (GO:0001222)	[CTBP1, LCOR, EZH2]	1.115464e-05
Double-Stranded DNA Binding (GO:0003690)	[BCL6, SATB1, HAND2, NFATC2, FOXP3, FOXP1]	4.115393e-05
Sequence-Specific Double-Stranded DNA Binding (GO:1990837)	[BCL6, HAND2, HAND1, NFATC2, FOXP3, FOXP1]	7.000979e-05
DNA Binding (GO:0003677)	[BCL6, SATB1, NFATC2, FOXP3, FOXP2, EZH2]	1.768160e-04

Figure 1.
Bioinformatic analysis of FOXP4: (A) protein–protein interaction (B) table of top 10 significant p-values for GO biological process (C) table of top 10 significant p-values for GO molecular function (D) KEGG pathway.

genes was found using the ShinyGO tool (http://bioinformatics.sdstate.edu/go/). SATB1, CTBP1, NFATC2, TBR1, FOXP4, FOXP3, FOXP2, GATAD2B, FOXP1, ZNF609, BCL6, HAND2, HAND1, LCOR, and EZH2 are involved in the regulation of DNA-templated transcription (GO:0006355) and transcription by RNA polymerase II (GO:0006357).

The following associated molecular function terms were also looked into: regulation of DNA-templated transcription (GO:0006355), regulation of transcription by RNA polymerase II (GO:0006357) for the proteins (SATB1, CTBP1, NFATC2, TBR1, FOXP4, FOXP3, FOXP2, GATAD2B, FOXP1, ZNF609, BCL6, HAND2, HAND1, LCOR, and EZH2).

2.2 A-MYB

Nuclear DNA-binding proteins known as myoblastosis (MYB) proteins function as transcriptional transactivators of several genes. Target genes that control cell cycle development, differentiation, and apoptosis are bound by them and transcriptionally stimulated. Mammals possess three distinct MYB paralogs: A-MYB, B-MYB, and C-MYB (correspondingly, MYBL1, MYBL2, and MYB in mice). Spermatocytes following meiotic entry have significant levels of MYBL1 transcription. Histological analysis indicates that the male infertility is caused by meiosis arrest at around the pachytene stage. Thus, it is plausible that MYBL1 functions as a crucial regulator of spermatocyte meiosis [22].

According to the study, a little alteration in transcription factor regulation upstream might have a cascade impact downstream in a regulatory network, resulting in hybrid male sterility. The relationship between the key transcription factor and hybrid male sterility was investigated using RNA sequencing, assay for transposase-accessible chromatin, and high-throughput sequencing analyses. It was discovered that the widespread misexpression in dzo was linked to spermatogenesis-related genes as well as somatic or progenitor genes. The male meiosis-specific master transcription factor MYBL1 motif was shown to be abundant on the promoters of downregulated pachytene spermatocyte genes in dzo, according to the study of transcription factor-binding motif enrichment. The MYBL1-binding motif's target genes were shown to be considerably downregulated in the dzo testis and linked to genes unique to meiosis [23].

Promoters, super enhancers, and endogenous retroviral enhancers that stimulate meiotic gene transcription are bound by A-MYB. A-MYB directly binds the promoters of less than 25% of meiotic genes, indicating that one or more additional transcription factors act downstream or in parallel to A-MYB to regulate meiotic gene transcription, even though A-MYB-mutant germ cells fail to express meiotic genes and arrest early in meiosis I [24].

2.2.1 The outcomes of bioinformatic analyses on MYBL-1

The protein–protein interaction network (**Figure 2A**) indicated the related genes including BCL7A, CREBBP, E2F2, E2F4, E2F5, FOXM1, LIN37, LIN52, LIN54, LIN9, LMO2, MYBL2, NCL, NFIB, RBBP4, RBL1, RBL2, TESMIN, TFDP1, and TFDP2.

CREBBP, LIN37, LMO2, LIN9, FOXM1, TFDP1, TFDP2, BCL7A, NFIB, E2F2, MYBL2, E2F4, E2F5, and MYBL1 all participated in the two biological processes

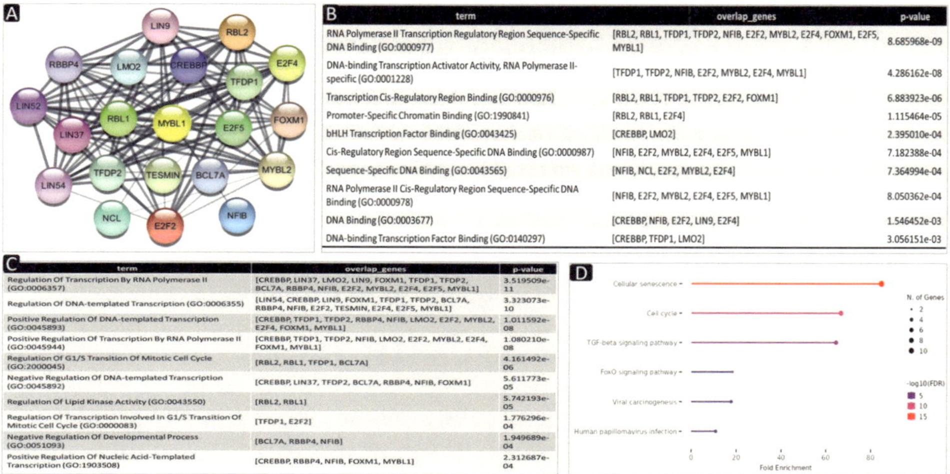

B

term	overlap_genes	p-value
RNA Polymerase II Transcription Regulatory Region Sequence-Specific DNA Binding (GO:0000977)	[RBL2, RBL1, TFDP1, TFDP2, NFIB, E2F2, MYBL2, E2F4, FOXM1, E2F5, MYBL1]	8.685968e-09
DNA-binding Transcription Activator Activity, RNA Polymerase II-specific (GO:0001228)	[TFDP1, TFDP2, NFIB, E2F2, MYBL2, E2F4, MYBL1]	4.286162e-08
Transcription Cis-Regulatory Region Binding (GO:0000976)	[RBL2, RBL1, TFDP1, TFDP2, E2F2, FOXM1]	6.883923e-06
Promoter-Specific Chromatin Binding (GO:1990841)	[RBL2, RBL1, E2F4]	1.115464e-05
bHLH Transcription Factor Binding (GO:0043425)	[CREBBP, LMO2]	2.395010e-04
Cis-Regulatory Region Sequence-Specific DNA Binding (GO:0000987)	[NFIB, E2F2, MYBL2, E2F4, E2F5, MYBL1]	7.182388e-04
Sequence-Specific DNA Binding (GO:0043565)	[NFIB, NCL, E2F2, MYBL2, E2F4]	7.364994e-04
RNA Polymerase II Cis-Regulatory Region Sequence-Specific DNA Binding (GO:0000978)	[NFIB, E2F2, MYBL2, E2F4, E2F5, MYBL1]	8.050362e-04
DNA Binding (GO:0003677)	[CREBBP, NFIB, E2F2, LIN9, E2F4]	1.546452e-03
DNA-binding Transcription Factor Binding (GO:0140297)	[CREBBP, TFDP1, LMO2]	3.056151e-03

C

term	overlap_genes	p-value
Regulation Of Transcription By RNA Polymerase II (GO:0006357)	[CREBBP, LIN37, LMO2, LIN9, FOXM1, TFDP1, TFDP2, BCL7A, RBBP4, NFIB, E2F2, MYBL2, E2F4, E2F5, MYBL1]	3.519509e-11
Regulation Of DNA-templated Transcription (GO:0006355)	[LIN54, CREBBP, LIN9, FOXM1, TFDP1, TFDP2, BCL7A, RBBP4, NFIB, E2F2, TESMIN, E2F4, E2F5, MYBL1]	3.323073e-10
Positive Regulation Of DNA-templated Transcription (GO:0045893)	[CREBBP, TFDP1, TFDP2, RBBP4, NFIB, LMO2, E2F2, MYBL2, E2F4, FOXM1, MYBL1]	1.011592e-08
Positive Regulation Of Transcription By RNA Polymerase II (GO:0045944)	[CREBBP, TFDP1, TFDP2, NFIB, LMO2, E2F2, MYBL2, E2F4, FOXM1, MYBL1]	1.080210e-08
Regulation Of G1/S Transition Of Mitotic Cell Cycle (GO:2000045)	[RBL2, RBL1, TFDP1, BCL7A]	4.161492e-06
Negative Regulation Of DNA-templated Transcription (GO:0045892)	[CREBBP, LIN37, TFDP2, BCL7A, RBBP4, NFIB, FOXM1]	5.611773e-05
Regulation Of Lipid Kinase Activity (GO:0043550)	[RBL2, RBL1]	5.742193e-05
Regulation Of Transcription Involved In G1/S Transition Of Mitotic Cell Cycle (GO:0000083)	[TFDP1, E2F2]	1.776296e-04
Negative Regulation Of Developmental Process (GO:0051093)	[BCL7A, RBBP4, NFIB]	1.949689e-04
Positive Regulation Of Nucleic Acid-Templated Transcription (GO:1903508)	[CREBBP, RBBP4, NFIB, FOXM1, MYBL1]	2.312687e-04

Figure 2.
Bioinformatic analysis of MYBL1: (A) protein–protein interaction (B) table of top 10 significant p-values for GO biological process (C) table of top 10 significant p-values for GO molecular function (D) KEGG pathway.

(regulation of transcription by RNA polymerase II (GO:0006357) and regulation of DNA-templated transcription (GO:0006355)). Furthermore, the linked molecular activities of this gene network include RNA Polymerase II transcription regulatory region sequence-specific DNA binding (GO:0000977), DNA-binding transcription activator activity, and RNA polymerase II-specific (GO:0001228).

2.3 TCFL5

Transcription factor-like 5 (TCFL5) stimulates the transcription of genes involved in meiosis, mRNA turnover, the generation of miR-34/449, meiotic exit, and spermiogenesis [24]. TCFL5 is found in the middle and late spermatocytes, beginning with stage V and progressing to the round spermatid at stage I, and has nuclear and cytoplasmic expression [25].

MYBL1 activates TCFL5 expression; in mice with defective MYBL1 testes, no TCFL5 protein is seen. Cleavage under targets and release using nuclease (CUT&RUN) for MYBL1 from FACS-purified male germ cells found MYBL1 occupancy around the TCFL5 gene promoter, supporting the same line of evidence. It is interesting to note that TCFL5 mutant testes had lower MYBL1 mRNA and protein levels. Interestingly, CUT&RUN for TCFL5 also recorded TCFL5 occupancy on the Mybl1 promoter, demonstrating that TCFL5 and MYBL1 positively feedback on each other's transcription [26]. Through high-throughput sequencing of TCFL5 ChIP-DNA and eCLIP-RNA, the study discovered that TCFL5 controls a group of genes involved in the formation of male germ cells both transcriptionally and posttranscriptionally. Additionally, the protein interacts with FXR1, a known RNA-binding protein that may be involved in coordinating TCFL5's transition and nucleus localization [25].

Galán *et al.* created two TCFL5 deletion mice models in order to study the function of TCFL5 in spermatogenesis [27]. Male mice that were part of the whole TCFL5-deficient mice paradigm had infertility due to Sox2 promoter-controlled CRE recombinase. This model's phenotypic investigation showed that TCFL5 is required for the progression of meiosis. Furthermore, the tamoxifen-conditional knockout mouse demonstrated that TCFL5, in addition to the previously mentioned functions, plays a role in spermiogenesis. According to gene expression research, the transcription factor TCFL5 may be able to regulate key spermatogenesis genes that are involved in mitosis, meiosis, and spermiogenesis. These findings highlight the critical function of TCFL5 in the advancement of spermatogenesis, which is necessary for the resolution of meiosis and the maturation of spermatids [27]. A further research conducted in 2023 on conservation in rhesus macaques indicates that TCFL5 is essential for placental animals' meiosis and spermiogenesis [24].

2.4 Gene network analysis related to TCFL5 and MYBL1

We chose to input these two genes into the program simultaneously in order to find any associated proteins because of the connection between them and their regulatory pathway. As in **Figure 3A**, ARFGEF1, BCL7A, GPR183, IL21, KIAA1549, LIN37, LIN52, LIN54, LIN9, LMO2, LRMP, MORN3, MOV10L1, NFIB, PDCD2, PPFIBP2, RBBP4, TMOD3, and ZFX were detected in the protein network. BCL7A, RBBP4, PDCD2, and TCFL5 are involved in the regulation of cell differentiation (GO:0045595). LIN37, BCL7A, RBBP4, NFIB, LMO2, LIN9, ZFX, MYBL1, and TCFL5 take place in the regulation of transcription by RNA polymerase II (GO:0006357). Moreover, MYBL1 and MOV10L1 can take part in male meiosis I (GO:0007141). Functional enrichment analysis demonstrated the biological process (BP) terms cis-regulatory region sequence-specific DNA binding (GO:0000987) and RNA polymerase II cis-regulatory region sequence-specific DNA binding (GO:0000978) ([NFIB, ZFX, MYBL1, TCFL5).

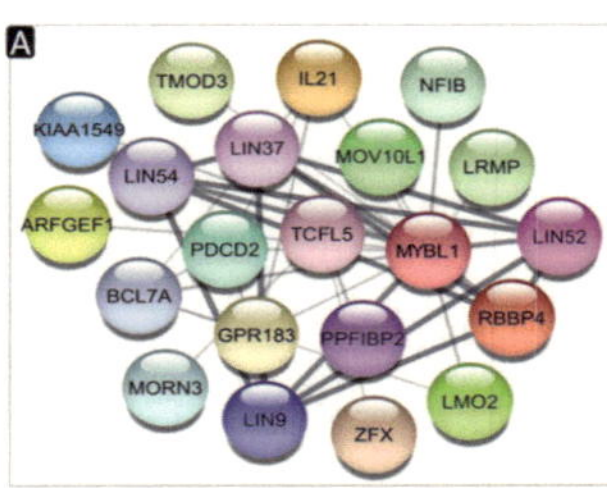

B

term	overlap_genes	p-value
Regulation Of Cell Differentiation (GO:0045595)	[BCL7A, RBBP4, PDCD2, TCFL5]	0.000044
Mature B Cell Differentiation Involved In Immune Response (GO:0002313)	[IL21, GPR183]	0.000081
Regulation Of Transcription By RNA Polymerase II (GO:0006357)	[LIN37, BCL7A, RBBP4, NFIB, LMO2, LIN9, ZFX, MYBL1, TCFL5]	0.000105
Male Meiosis I (GO:0007141)	[MOV10L1, MYBL1]	0.000125
T-helper Cell Differentiation (GO:0042093)	[IL21, GPR183]	0.000159
Negative Regulation Of Developmental Process (GO:0051093)	[BCL7A, RBBP4, NFIB]	0.000195
Meiosis I (GO:0007127)	[MOV10L1, MYBL1]	0.000336
Positive Regulation Of B Cell Proliferation (GO:0030890)	[IL21, GPR183]	0.000419
Regulation Of DNA-templated Transcription (GO:0006355)	[LIN54, BCL7A, RBBP4, NFIB, LIN9, ZFX, MYBL1, TCFL5]	0.000464
Positive Regulation Of Stem Cell Population Maintenance (GO:1902459)	[BCL7A, RBBP4]	0.000800

C

term	overlap_genes	p-value
Chromatin DNA Binding (GO:0031490)	[RBBP4, ZFX]	0.003395
DNA-binding Transcription Activator Activity, RNA Polymerase II-specific (GO:0001228)	[NFIB, ZFX, MYBL1]	0.005505
Minor Groove Of Adenine-Thymine-Rich DNA Binding (GO:0003680)	[LIN54]	0.006284
Cadherin Binding Involved In Cell-Cell Adhesion (GO:0098641)	[TMOD3]	0.017708
bHLH Transcription Factor Binding (GO:0043425)	[LMO2]	0.022859
Protein Kinase A Regulatory Subunit Binding (GO:0034237)	[ARFGEF1]	0.022859
Cis-Regulatory Region Sequence-Specific DNA Binding (GO:0000987)	[NFIB, ZFX, MYBL1, TCFL5]	0.025584
RNA Polymerase II Cis-Regulatory Region Sequence-Specific DNA Binding (GO:0000978)	[NFIB, ZFX, MYBL1, TCFL5]	0.027443
DNA Secondary Structure Binding (GO:0000217)	[LIN54]	0.034101
Nucleosomal DNA Binding (GO:0031492)	[RBBP4]	0.035117

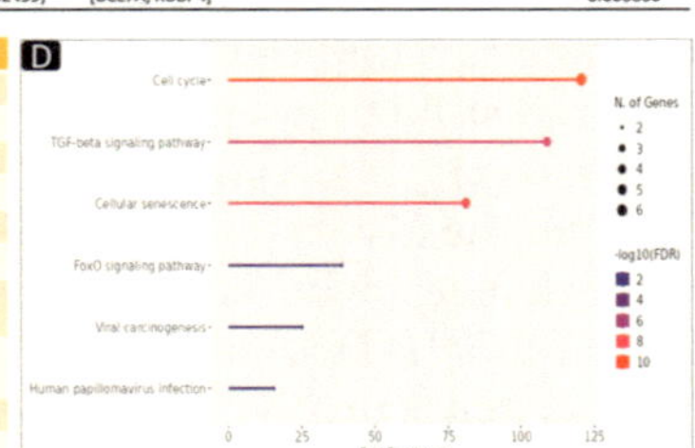

Figure 3.
Bioinformatic analysis of TCFL5: (A) protein–protein interaction (B) table of top 10 significant p-values for GO biological process (C) table of top 10 significant p-values for GO molecular function (D) KEGG pathway.

2.5 TCF3

The E protein E2A, a member of the basic helix–loop–helix (b-HLH) transcription factor family, is encoded by transcription factor 3 (TCF3). Two unique isoforms, E47 and E12, are produced by alternative splicing; they only vary in exon 18 pertaining to the b-HLH domain [28]. To carry out their tissue- or cell type-specific tasks, TCF3 proteins combine with other b-HLH proteins to create homodimers or heterodimers. The TCF3 protein is dysregulated in a number of malignancies, including lymphoma, pancreatic cancer, breast cancer, colorectal cancer, and prostate cancer. It was first shown to play especially significant functions during lymphocyte development. Numerous investigations have demonstrated that the TCF3 protein functions as an oncoprotein or a tumor suppressor [29].

The co-occupancy of TCF3 with Oct4, Sox2, and Nanog on embryonic stem cell (ESC) chromatin indicated that TCF3 has been suggested to play an integral role in a poorly understood mechanism underlying wingless-related integration site-dependent stimulation of mouse ESC self-renewal of mouse ESCs [30].

According to a recent study, GFRA1-positive spermatogonia were the main source of TCF3 expression, and TCF3 expression was upregulated by the epidermal growth factor (EGF). Interestingly, TCF3 inhibited human SSC apoptosis while promoting SSC proliferation and DNA synthesis. TCF3 protein controlled the transcription of many genes, including WNT2B, TGFB3, CCN4, MEGF6, and PODXL, according to RNA sequencing and chromatin immunoprecipitation (ChIP) experiments. PODXL silencing, on the other hand, reduced the stem cell activity of SSCs. Furthermore, in patients with spermatogenesis failure, the amount of TCF3 protein was significantly lower than in those with obstructive azoospermia who had normal spermatogenesis. Altogether, our findings suggested that TCF3 regulates PODXL-mediated human SSC proliferation and apoptosis [31].

2.5.1 Bioinformatics analysis of TCF3

The gene was uploaded in Cytoscape software, and the related proteins are demonstrated in **Figure 4A**. (TCF3, CBFA2T3, CBFB, GATA1, GFI1B, ID1, ID2, LDB1, LDB2, LMO1, LMO2, LYL1, MYOD1, RUNX1, SSBP2, SSBP3, TAL1, TAL2, TCF12, TLX1, and ZFPM1). The related biological processes after enrichment were: DNA-binding transcription factor binding (GO:0140297), RNA polymerase II-specific DNA-binding transcription factor binding (GO:0061629), cis-regulatory region sequence-specific DNA binding (GO:0000987), RNA polymerase II cis-regulatory region sequence-specific DNA binding (GO:0000978), RNA polymerase II transcription regulatory region sequence-specific DNA binding (GO:0000977), and DNA-binding transcription activator activity, RNA polymerase II-specific (GO:0001228).

Moreover, the molecular functions of the gene set were: regulation of transcription by RNA polymerase II (GO:0006357), positive regulation of DNA-templated transcription (GO:0045893), regulation of DNA-templated transcription (GO:0006355), positive regulation of transcription by RNA polymerase II (GO:0045944), negative regulation of DNA-templated transcription (GO:0045892), and negative regulation of transcription by RNA polymerase II (GO:0000122).

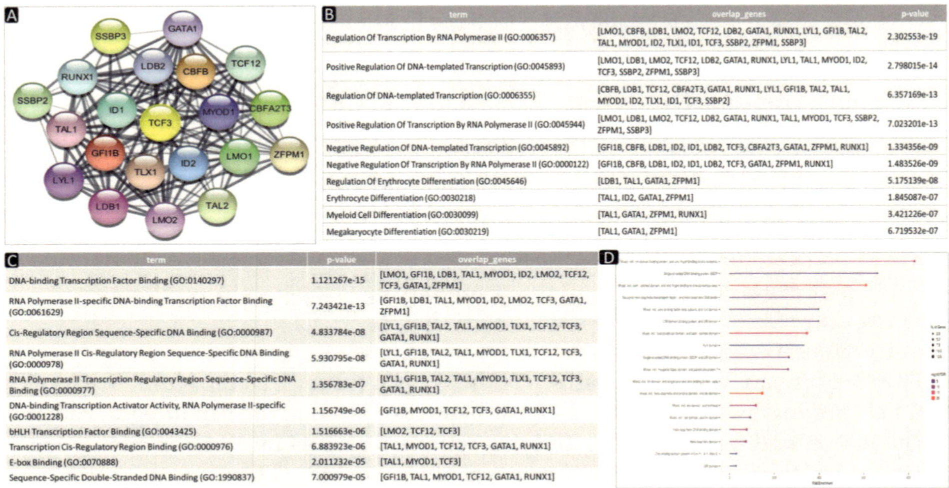

B

term	overlap_genes	p-value
Regulation Of Transcription By RNA Polymerase II (GO:0006357)	[LMO1, CBFB, LDB1, LMO2, TCF12, LDB2, GATA1, RUNX1, LYL1, GFI1B, TAL2, TAL1, MYOD1, ID2, TLX1, ID1, TCF3, SSBP2, ZFPM1, SSBP3]	2.302553e-19
Positive Regulation Of DNA-templated Transcription (GO:0045893)	[LMO1, LDB1, LMO2, TCF12, LDB2, GATA1, RUNX1, LYL1, TAL1, MYOD1, ID2, TCF3, SSBP2, ZFPM1, SSBP3]	2.798015e-14
Regulation Of DNA-templated Transcription (GO:0006355)	[CBFB, LDB1, TCF12, CBFA2T3, GATA1, RUNX1, LYL1, GFI1B, TAL2, TAL1, MYOD1, ID2, TLX1, ID1, TCF3, SSBP2]	6.357169e-13
Positive Regulation Of Transcription By RNA Polymerase II (GO:0045944)	[LMO1, LDB1, LMO2, TCF12, LDB2, GATA1, RUNX1, TAL1, MYOD1, TCF3, SSBP2, ZFPM1, SSBP3]	7.023201e-13
Negative Regulation Of DNA-templated Transcription (GO:0045892)	[GFI1B, CBFB, LDB1, ID2, ID1, LDB2, TCF3, CBFA2T3, GATA1, ZFPM1, RUNX1]	1.334356e-09
Negative Regulation Of Transcription By RNA Polymerase II (GO:0000122)	[GFI1B, CBFB, LDB1, ID2, ID1, LDB2, TCF3, GATA1, ZFPM1, RUNX1]	1.483526e-09
Regulation Of Erythrocyte Differentiation (GO:0045646)	[LDB1, TAL1, GATA1, ZFPM1]	5.175139e-08
Erythrocyte Differentiation (GO:0030218)	[TAL1, ID2, GATA1, ZFPM1]	1.845087e-07
Myeloid Cell Differentiation (GO:0030099)	[TAL1, GATA1, ZFPM1, RUNX1]	3.421226e-07
Megakaryocyte Differentiation (GO:0030219)	[TAL1, GATA1, ZFPM1]	6.719532e-07

C

term	p-value	overlap_genes
DNA-binding Transcription Factor Binding (GO:0140297)	1.121267e-15	[LMO1, GFI1B, LDB1, TAL1, MYOD1, ID2, LMO2, TCF12, TCF3, GATA1, ZFPM1]
RNA Polymerase II-specific DNA-binding Transcription Factor Binding (GO:0061629)	7.243421e-13	[GFI1B, LDB1, TAL1, MYOD1, ID2, LMO2, TCF3, GATA1, ZFPM1]
Cis-Regulatory Region Sequence-Specific DNA Binding (GO:0000987)	4.833784e-08	[LYL1, GFI1B, TAL2, TAL1, MYOD1, TLX1, TCF12, TCF3, GATA1, RUNX1]
RNA Polymerase II Cis-Regulatory Region Sequence-Specific DNA Binding (GO:0000978)	5.930795e-08	[LYL1, GFI1B, TAL2, TAL1, MYOD1, TLX1, TCF12, TCF3, GATA1, RUNX1]
RNA Polymerase II Transcription Regulatory Region Sequence-Specific DNA Binding (GO:0000977)	1.356783e-07	[LYL1, GFI1B, TAL2, TAL1, MYOD1, TLX1, TCF12, TCF3, GATA1, RUNX1]
DNA-binding Transcription Activator Activity, RNA Polymerase II-specific (GO:0001228)	1.156749e-06	[GFI1B, MYOD1, TCF12, TCF3, GATA1, RUNX1]
bHLH Transcription Factor Binding (GO:0043425)	1.516663e-06	[LMO2, TCF12, TCF3]
Transcription Cis-Regulatory Region Binding (GO:0000976)	6.883923e-06	[TAL1, MYOD1, TCF12, TCF3, GATA1, RUNX1]
E-box Binding (GO:0070888)	2.011232e-05	[TAL1, MYOD1, TCF3]
Sequence-Specific Double-Stranded DNA Binding (GO:1990837)	7.000979e-05	[GFI1B, TAL1, MYOD1, TCF12, GATA1, RUNX1]

Figure 4.
Bioinformatic analysis of TCF3: (A) protein–protein interaction (B) table of top 10 significant p-values for GO biological process (C) table of top 10 significant p-values for GO molecular function (D) KEGG pathway.

2.6 E4F1

The human genome's 16p13.3 region contains the E4F transcription factor 1 (E4F1) gene, which codes for a zinc finger (ZF) protein belonging to the GLI-Kruppel family. This protein is widely acknowledged as a transcription factor and is produced in both humans and mice [32]. The E4F1 protein was initially discovered as a transcriptional regulator of the E4 viral gene, which is hijacked during adenoviral infection by the viral oncoprotein E1A [33].

As a multifunctional transcription factor, E4F1 plays an essential role in cell fate decisions and interacts with Retinoblastoma protein (pRB), a regulator of Sertoli cells [34]. Due to the mortality of the preimplantation embryos in E4F1 knockout mice, the E4F1 is thought as a necessary gene for early embryonic development [32]. A study that used a mouse model with a conditional knockout of the E4F1 gene showed that E4F1 deletion affected Sertoli cell proliferation, which in turn increased germ cell death and decreased testis size and fertility in order to avoid this embryonic mortality [34].

Yan et al. examined undifferentiated spermatogonia's chromatin accessibility at the single-cell level, and they discovered 37 positive TF regulators that could play a part in determining the fates of SSCs [35]. The whole undifferentiated spermatogonial pool gradually disappears when the transcription factor E4F1, which is expressed in spermatogonia, is conditionally deleted in mouse germ cells. E4F1 functions as a critical modulator of mitochondrial activity, as demonstrated by single-cell RNA-seq study of normal and E4f1-deficient spermatogonia. Ndufs5, Cox7a2, Cox6c, and Dnajc19 are among the genes whose promotor E4F1 binds to and encodes components of the mitochondrial respiratory chain. Undifferentiated spermatogonia were progressively lost as a result of cell cycle arrest and increased apoptosis brought on by aberrant mitochondrial morphology and abnormalities in fatty acid metabolism

DOI: http://dx.doi.org/10.5772/intechopen.1004178

induced by loss of E4F1 activity. P53 deletion in E4F1-deficient germ cells did not correct the abnormalities in SSC maintenance; rather, it merely momentarily stopped spermatogonial loss [35].

2.6.1 Bioinformatic analysis of E4F1

The network was ultimately extended once this gene was added to the Cytoscape program.

This gene was observed in connection with BMI1, BRAT1, CBX2. CBX4. CBX6, CBX7, CDKN2A, CITED2, COMMD3-BMI1, FAAP100, LRRC14, PHC1, PHC2, RASSF1, SCMH1, TELO2, TP53, WDR24, ZBTB17, and ZGPAT (**Figure 5A**).

As we are able to perceive, E4F1 with TP53, ZBTB17, and CITED2 are involved in negative regulation of DNA-templated transcription (GO:0045892), regulation of cell cycle (GO:0051726), negative regulation of transcription by RNA polymerase II (GO:0000122), and regulation of transcription by RNA polymerase II (GO:0006357).

Also, the molecular function terms including: RNA polymerase II-specific DNA-binding transcription factor binding (GO:0061629) for CDKN2A, E4F1 and TP53, DNA-binding transcription repressor activity, RNA polymerase II-specific (GO:0001227) for ZGPAT, E4F1 and TP53, DNA-binding transcription factor binding (GO:0140297) and DNA-binding transcription activator activity, RNA polymerase II-specific (GO:0001228) for ZBTB17, E4F1 and TP53. KEGG pathway shows that these genes can be involved in signaling pathways regulating pluripotent stem cells.

Our attention was drawn to Cbp/P300-Interacting Transactivator with Glu/Asp-Rich Carboxy-Terminal Domain 2 (CITED2), a transcription factor that may

A

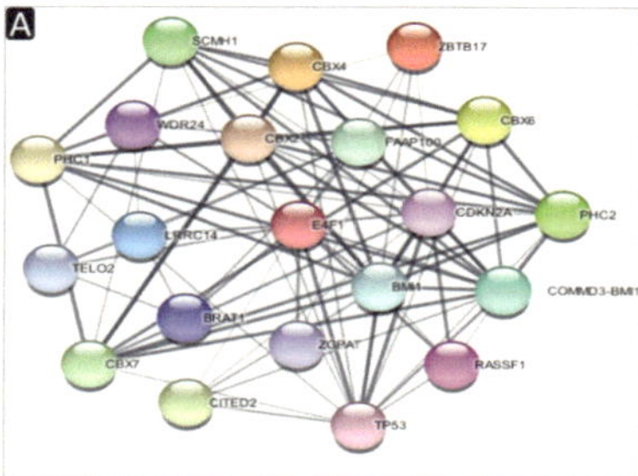

B

term	overlap_genes	p-value
Negative Regulation Of DNA-templated Transcription (GO:0045892)	[CBX7, ZBTB17, CBX6, ZGPAT, CBX4, CDKN2A, CITED2, CBX2, SCMH1, E4F1, BMI1, TP53]	5.898480e-11
Negative Regulation Of Transcription By RNA Polymerase II (GO:0000122)	[CBX7, ZBTB17, CBX6, ZGPAT, CBX4, CITED2, CBX2, E4F1, BMI1, TP53]	1.483526e-09
Negative Regulation Of Nucleic Acid-Templated Transcription (GO:1903507)	[ZGPAT, CBX4, CDKN2A, CITED2, SCMH1, TP53]	5.514487e-06
Regulation Of Cell Cycle (GO:0051726)	[ZBTB17, CDKN2A, CITED2, E4F1, TP53]	1.188074e-05
Regulation Of Transcription By RNA Polymerase II (GO:0006357)	[CBX7, ZBTB17, CBX6, ZGPAT, CBX4, CITED2, CBX2, E4F1, BMI1, TP53]	1.384366e-05
Small GTPase Mediated Signal Transduction (GO:0007264)	[RASSF1, CDKN2A, TP53]	2.112447e-04
Positive Regulation Of TORC1 Signaling (GO:1904263)	[WDR24, TELO2]	3.627682e-04
Ras Protein Signal Transduction (GO:0007265)	[RASSF1, CDKN2A, TP53]	4.420414e-04
Regulation Of Fibroblast Proliferation (GO:0048145)	[BMI1, TP53]	1.056927e-03
Positive Regulation Of TOR Signaling (GO:0032008)	[WDR24, TELO2]	1.103204e-03

C

term	overlap_genes	p-value
Core Promoter Sequence-Specific DNA Binding (GO:0001046)	[ZBTB17, TP53]	0.000612
Promoter-Specific Chromatin Binding (GO:1990841)	[BMI1, TP53]	0.000881
RNA Polymerase II-specific DNA-binding Transcription Factor Binding (GO:0061629)	[CDKN2A, E4F1, TP53]	0.001671
DNA-binding Transcription Repressor Activity, RNA Polymerase II-specific (GO:0001227)	[ZGPAT, E4F1, TP53]	0.002004
Methylation-Dependent Protein Binding (GO:0140034)	[CBX6, CBX2]	0.002294
Methylated Histone Binding (GO:0035064)	[CBX6, CBX2]	0.002498
DNA-binding Transcription Factor Binding (GO:0140297)	[ZBTB17, E4F1, TP53]	0.003056
Zinc Ion Binding (GO:0008270)	[RASSF1, BMI1, TP53]	0.005203
LBD Domain Binding (GO:0050693)	[CITED2]	0.005239
DNA-binding Transcription Activator Activity, RNA Polymerase II-specific (GO:0001228)	[ZBTB17, E4F1, TP53]	0.005505

D

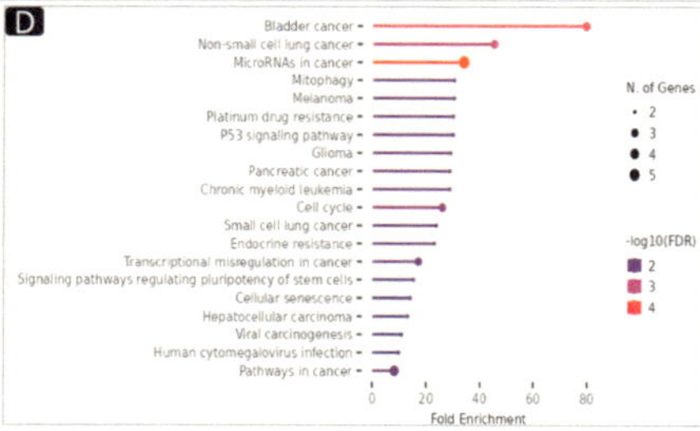

Figure 5.
Bioinformatic analysis of E4F1: (A) protein–protein interaction (B) table of top 10 significant p-values for GO biological process (C) table of top 10 significant p-values for GO molecular function (D) KEGG pathway.

be involved in the process of spermatogenesis, by further gene network search and analysis. We decided to identify this gene for further research and to elucidate its function in spermatogenesis after reviewing the literature.

2.7 CITED2

CITED2 (a key member of the CITED family), a transcription cofactor, engages with various transcription factors and cofactors, playing crucial roles in fundamental cellular processes such as proliferation, apoptosis, differentiation, migration, and autophagy. Its interacting partners, including LIM homeobox 2, transcription factor AP-2, SMAD2/3, peroxisome proliferator-activated receptor γ, estrogen receptor, MYC, Nucleolin, and p300/CBP, actively regulate downstream gene expression, contributing significantly to the mentioned cellular processes. Recent findings have underscored the indispensable function of CITED2 in both embryonic and adult tissue stem cells, encompassing hematopoietic stem cells and tendon-derived stem/progenitor cells [36, 37]. In the study conducted by Buaas et al. in 2009, it was determined that the transcription cofactor CITED2 plays a crucial role in the processes of sex determination and the early development of gonads [38]. According to the findings of the study conducted by Fang et al. it is proposed that the measurement of CITED2 protein expression in cumulus cells can be a valuable tool for assessing embryo quality and predicting the probability of a successful pregnancy, particularly within the specific context of an *in vitro* fertilization (IVF) cycle. This implies that monitoring the levels of CITED2 in cumulus cells may offer clinicians and researchers a reliable indicator to evaluate the viability of embryos and anticipate the likelihood of a positive pregnancy outcome during IVF procedures [39].

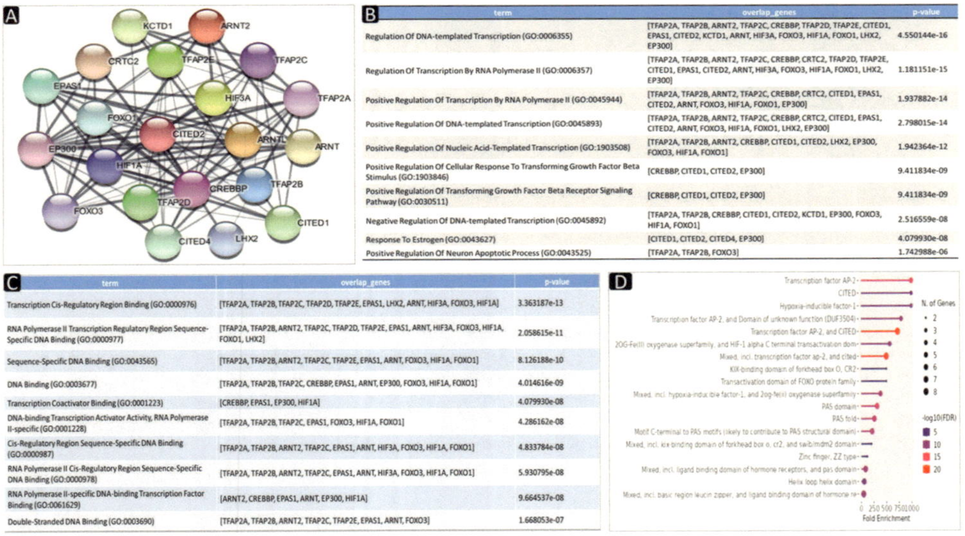

Figure 6.
Bioinformatic analysis of CITED2: (A) protein–protein interaction (B) table of top 10 significant p-values for GO biological process (C) table of top 10 significant p-values for GO molecular function (D) KEGG pathway.

DOI: http://dx.doi.org/10.5772/intechopen.1004178

2.7.1 Bioinformatic analysis of CITED2

Similar to the previous genes, bioinformatic analyses were conducted for this gene, and the following results were obtained. With regard to **Figure 6A**, CITED2 is in interaction with ARNT, ARNT2, ARNTL, CITED1, CITED4, CREBBP, CRTC2, EP300, EPAS1, FOXO1, FOXO3, HIF1A, HIF3A, KCTD1, LHX2, TFAP2A, TFAP2B, TFAP2C, TFAP2D, and TFAP2E.

TFAP2A, TFAP2B, ARNT2, TFAP2C, CREBBP, CRTC2, TFAP2D, TFAP2E, CITED1, EPAS1, CITED2, ARNT, HIF3A, FOXO3, HIF1A, FOXO1, LHX2, and EP300 are involved in the regulation of DNA-templated transcription (GO:0006355) and transcription by RNA polymerase II (GO:0006357).

3. Conclusion

Spermatogenesis is an intricate process involving numerous gene regulations and transcription factors. Spermatogonial stem cells (SSCs) play a pivotal role in initiating and sustaining spermatogenesis. Transcription factors, by regulating gene expression, activate SSCs, leading to the formation of sperm. These recent discoveries have contributed to a better understanding of spermatogenesis, enabling us to anticipate molecular connections and functional characteristics of proteins involved in SSC activation for sperm formation. This book chapter highlights the crucial role of transcription factors in the intricate regulation of gene expression in SSCs. Our bioinformatics research indicates that the CITED2 gene could play a valuable role in spermatogenesis. Existing studies suggest the potential significance and essential nature of this protein. Nevertheless, further research is necessary to validate this claim. A comprehensive analysis of current research indicates that these transcription factors might serve as biomarkers for gene therapy and other treatments related to male infertility.

Acknowledgements

This study was funded by Center for International Scientific Studies and Collaboration (CISSC) and Amol University of Special Modern Technologies.

Conflict of interest

The authors declare no conflict of interest.

Abbreviations

SSCs	spermatogonia stem cells
TFs	transcription factors
E4F1	E4F transcription factor 1
FOXP4	Forkhead box protein P4
A-MYB	MYB Proto-Oncogene Like 1
TCFL5	transcription factor like 5

TCF3	transcription factor 3
CITED2	Cbp/P300-Interacting Transactivator with Glu/Asp-Rich Carboxy-Terminal Domain 2
ChIP	chromatin immunoprecipitation
RSTs	round spermatids
Plzf	promyelocytic leukemia zinc finger protein

Author details

Zahra Hasani Mahforoozmahalleh[1] and Hossein Azizi[2]*

1 Department of Microbial Biotechnology, Amol University of Special Modern Technologies, Amol, Iran

2 Faculty of Biotechnology, Amol University of Special Modern Technologies, Amol, Iran

*Address all correspondence to: h.azizi@ausmt.ac.ir

References

[1] Niazi Tabar A, Azizi H, Hashemi Karoii D, Skutella T. Testicular localization and potential function of vimentin positive cells during spermatogonial differentiation stages. Animals. 2022;**12**(3):268

[2] Huang Y et al. Single-cell multi-omics sequencing of human spermatogenesis reveals a DNA demethylation event associated with male meiotic recombination. Nature Cell Biology. 2023;**25**(10):1520-1534

[3] Liu S et al. Identification of hub genes associated with spermatogenesis by bioinformatics analysis. Scientific Reports. 2023;**13**(1):18435

[4] Ali W et al. A new insight of cadmium-induced cellular evidence of autophagic-associated spermiophagy during spermatogenesis. Environmental Science and Pollution Research International. 2023;**30**(45):101064-101074

[5] Chao H-H et al. Comprehensive review on the positive and negative effects of various important regulators on male spermatogenesis and fertility. Frontiers in Nutrition. 2023;**9**:1063510

[6] Danial Hashemi K, Hossein A. Undifferentiated and differentiated Spermatogonial stem cells. In: Leisheng ZP, editor. Advances in Pluripotent Stem Cells. London, UK, Rijeka: IntechOpen; 2023. Ch. 10

[7] Hashemi Karoii D, Azizi H. Functions and mechanism of noncoding RNA in regulation and differentiation of male mammalian reproduction. Cell Biochemistry and Function. 2023;**41**(7):767-778

[8] Gao C et al. The functions of FOXP transcription factors and their regulation by post-translational modifications. Biochimica et Biophysica Acta (BBA)-Gene Regulatory Mechanisms. 2023;**1866**:194992

[9] Shi K et al. Integrated bioinformatics analysis of the transcription factor-mediated gene regulatory networks in the formation of spermatogonial stem cells. Frontiers in Physiology. 2022;**13**:949486

[10] Luo SW et al. FOXP4 promotes proliferation of human spermatogonial stem cells. Asian Journal of Andrology. 2023;**25**(3):322-330

[11] Zhang D et al. The transcription factor SOX30 is a key regulator of mouse spermiogenesis. Development. 2018;**145**(11):dev164723

[12] Amirian M, Azizi H, Hashemi Karoii D, Skutella T. VASA protein and gene expression analysis of human non-obstructive azoospermia and normal by immunohistochemistry, immunocytochemistry, and bioinformatics analysis. Scientific Reports. 2022;**12**(1):17259

[13] Azizi H, Hashemi Karoii D, Skutella T. Whole exome sequencing and in silico analysis of human sertoli in patients with non-obstructive azoospermia. International Journal of Molecular Sciences. 2022;**23**(20):12570

[14] Hashemi Karoii D, Azizi H. OCT4 protein and gene expression analysis in the differentiation of spermatogonia stem cells into neurons by immunohistochemistry, immunocytochemistry, and bioinformatics analysis. Stem Cell Reviews and Reports. 2023;**19**:1-17

[15] Hashemi Karoii D, Azizi H, Skutella T. Altered G-protein transduction protein gene expression in the testis of infertile patients with nonobstructive azoospermia. DNA and Cell Biology. 2023;**42**(10):617-637

[16] Hashemi Karoii D, Azizi H, Skutella T. Microarray and in silico analysis of DNA repair genes between human testis of patients with nonobstructive azoospermia and normal cells. Cell Biochemistry and Function. 2022;**40**(8):865-879

[17] Karoii DH, Azizi H, Amirian M. Signaling pathways and protein–protein interaction of vimentin in invasive and migration cells: A review. Cellular Reprogramming. 2022;**24**(4):165-174

[18] Xu J et al. The role of Forkhead box family in bone metabolism and diseases. Frontiers in Pharmacology. 2021;**12**:772237

[19] Takahashi K, Liu F-C, Hirokawa K, Takahashi H. Expression of Foxp4 in the developing and adult rat forebrain. Journal of Neuroscience Research. 2008;**86**(14):3106-3116

[20] Shi J et al. FOXP4 promotes laryngeal squamous cell carcinoma progression through directly targeting LEF-1. Molecular Medicine Reports. 2021;**24**(6):831

[21] Matsumoto T et al. FOXP4 inhibits squamous differentiation of atypical cells in cervical intraepithelial neoplasia via an ELF3-dependent pathway. Cancer Science. 2022;**113**(10):3376-3389

[22] Bolcun-Filas E et al. A-MYB (MYBL1) transcription factor is a master regulator of male meiosis. Development. 2011;**138**(15):3319-3330

[23] Zhang G-W et al. Evaluation of MYBL1 as the master regulator for pachytene spermatocyte genes dysregulated in interspecific hybrid dzo. Journal of Dairy Science. 2023;**106**(6):4366-4379

[24] Cecchini K et al. The transcription factor TCFL5 responds to A-MYB to elaborate the male meiotic program in mice. Reproduction. 2023;**165**(2):183-196

[25] Xu W et al. Transcription factor-like 5 is a potential DNA-and RNA-binding protein essential for maintaining male fertility in mice. Journal of Cell Science. 2022;**135**(3):jcs259036

[26] Säflund M, Özata DM. The MYBL1/TCFL5 transcription network: Two collaborative factors with central role in male meiosis. Biochemical Society Transactions. 2023;**51**:2163-2172

[27] Galán-Martínez J et al. TCFL5 deficiency impairs the pachytene to diplotene transition during spermatogenesis in the mouse. Scientific Reports. 2022;**12**(1):10956

[28] Escherich C et al. Identification of TCF3 germline variants in pediatric B-cell acute lymphoblastic leukemia. Blood Advances. 2023;**7**(10):2177-2180

[29] Gui T et al. TCF3 is epigenetically silenced by EZH2 and DNMT3B and functions as a tumor suppressor in endometrial cancer. Cell Death & Differentiation. 2021;**28**(12):3316-3328

[30] Yi F et al. Opposing effects of Tcf3 and Tcf1 control Wnt stimulation of embryonic stem cell self-renewal. Nature Cell Biology. 2011;**13**(7):762-770

[31] Zhou D et al. TCF3 regulates the proliferation and apoptosis of human spermatogonial stem cells by targeting PODXL. Frontiers in Cell and Developmental Biology. 2021;**9**:695545

[32] Sun S et al. Transcription factor E4F1 as a regulator of cell life and disease progression. Science Advances. 2023;**9**(39):eadh1991

[33] Chua BH, Anuar NZ, Ferry L, Domrane C, Wittek A, Mukundan VT, et al. E4F1 and ZNF148 are transcriptional activators of the– 57A> C and wild-type TERT promoter. Genome Research. 1 Nov 2023;**33**(11):1893-1905

[34] Yan R-G, Yang Q-L, Yang Q-E. E4 transcription factor 1 (E4F1) regulates sertoli cell proliferation and fertility in mice. Animals. 2020;**10**(9):1691

[35] Yan R-G et al. Transcription factor E4F1 dictates spermatogonial stem cell fate decisions by regulating mitochondrial functions and cell cycle progression. Cell & Bioscience. 2023;**13**(1):177

[36] An B, Ji X, Gong Y. Role of CITED2 in stem cells and cancer. Oncology Letters. 2020;**20**(4):1-1

[37] Zheng SQ et al. Genetic analysis of the CITED2 gene promoter in isolated and sporadic congenital ventricular septal defects. Journal of Cellular and Molecular Medicine. 2021;**25**(4):2254-2261

[38] Buaas FW, Val P, Swain A. The transcription co-factor CITED2 functions during sex determination and early gonad development. Human Molecular Genetics. 2009;**18**(16):2989-3001

[39] Fang Y, Shang W, Wei D-L, Zeng S-M. Cited2 protein level in cumulus cells is a biomarker for human embryo quality and pregnancy outcome in one in vitro fertilization cycle. Fertility and Sterility. 2016;**105**(5):1351-1359. e4

Section 2

Assistant Reproductive Technology

Chapter 2

Perspective Chapter: Application of Abnormally Fertilized Eggs and the Associated Clinical Outcomes

Bin Wu, Xue Feng, Suzhen Lu and Timothy J. Gelety

Abstract

Human IVF laboratory often shows many abnormal fertilization eggs, such as no pronucleus (0PN), monopronucleus (1PN), three pronuclei (3PN) or multi-pronuclei (mPN) zygotes and these abnormal eggs are usually not used and typically discarded. Recent researches have showed that further evaluation on these abnormal eggs may provide some hope for aged infertile couples to have babies in their family. Our study showed that 0PN eggs may be rescued by introcytoplasmic sperm injection (ICSI). The 1PN zygotes should be cultured to observe their cleavage and blastocyst formation until Day 3 to Day 6. Selecting normal chromosomal embryo transfer may have healthy baby birth. Although most of 3PN embryos are genetically abnormal chromosomal composition, a small portion of 3PN embryos may develop to blastocyst with normal chromosomal composition. Also, those embryos derived from 3PN, especially by ICSI, have more possibility for self-correction to become normal euploid embryos. The microsurgically removing 1PN from 3PN zygotes may artificially correct this abnormal fertilization. After one PN removal, the formed blastocyst may be screened by the PGT for embryo transfer in rare embryo patients to achieve pregnancy and delivery of a healthy newborn. Based on no obvious difference of ooplasm between normal fertilized 2PN and 3PN zygotes, the cytoplasm of 3PN zygotes may be used to supplement the aged woman poor oocytes to improve embryo quality. Transferring partial cytoplasm from 3PN zygote to the fertilized 2PN zygotes of aged woman may promote the receipt embryo to develop blastocysts. This partial ooplasmic transfer does not change the aging woman genetic composition and the woman embryos still keep her with her husband genetic genes in the cell nucleus. However, the baby born with this technique might appear epigenetics because the mixed mitochondrial DNA would be passed on to all future generations.

Keywords: human IVF, abnormal fertilization, 0PN, 1PN, 3PN, embryo selection, self-correction, enucleate, cytoplasmic transfer

1. Introduction

Assisted reproductive technology has been widely used for the treatment of infertile couples to realize their dream of having a baby. Currently, this technology mainly contains *in vitro* fertilization (IVF) and its related procedures—intracytoplasmic sperm injection (ICSI). The key point of this technique is that the patient needs to be administered with some medicine to stimulate the ovaries to produce more oocytes during one reproductive period cycle and the oocytes on ovaries will be retrieved. After oocyte retrieval, insemination is performed with the use of conventional IVF or intracytoplasmic sperm injection (ICSI) under the laboratory condition. In all IVF laboratories, normal fertilization is determined by the presence of two pronuclei (2PN) associated with extrusion of the second polar body at 16–18 h post-insemination. Those zygotes displaying successful extrusion of the second polar body and two even pronuclei (2PN) are considered to be "normally fertilized," and they are cultured further to observe embryo quality and development. An absence of either 2PN or the second polar body indicates fertilization failure in which eggs show zero pronucleus (0PN). In the other way, those zygotes showing a single or more than 2PN are considered to be "abnormally fertilized." Sometimes, three or more PN are visible in the cytoplasm of the oocyte at the fertilization check, which indicates either polyspermy or failure of extrusion of the second polar body. Occasionally, only 1PN is seen in the cytoplasm of the oocyte, and this is generally attributed to oocyte parthenogenetic activation, irregular pronuclear formation resulting from asynchrony of pronuclear appearance, or possibly male and female pronuclear fusion. Zygotes with one PN are thought to be at a higher risk of being haploid, and the transfer of these embryos is expected to result in an implantation failure [1]. Even though the zygotes deriving from abnormally fertilized oocytes are capable of normal *in vitro* development, they are usually discarded because of a higher risk for abnormal ploidy constitution (i.e., haploidy, triploidy, or tetraploidy) [2]. Overall, $3 \sim 10\%$ of inseminated oocytes fertilize abnormally, and the embryos deriving from them are typically discarded in the absence of a reliable approach to monitor their genetic risk for ploidy defects [3]. However, recently, some research analysis has showed that abnormally fertilized oocytes have given development potential and can produce healthy live births [4–7]. In order to evaluate character of abnormal fertilization eggs, this chapter will concentrate on discussing the application of unfertilized and abnormal fertilization oocytes.

2. IVF failure and abnormal fertilization

Currently, human IVF technique is widely used in treatment of infertile couples to realize their family desire to have a child. However, *in vitro* fertilization is a complex process that requires many steps, which involves the combination for normal sperm with normal oocyte. As we know, all cells in the human body consist of two copies of each chromosome in 23 pairs of chromosome: one set is from the mother, and another set is from the father. The gametes (sperm and oocytes) contain only one copy of the chromosomes. Thus, both oocyte and sperm are haploid. During oocyte maturation, the oocyte needs to go through meiosis and to repel its excess chromosomes by releasing polar bodies to form metaphase II (MII) oocytes. The first polar body is extruded around the time of ovulation or oocyte retrieval. The presence of the first polar body indicates the oocyte mature situation that has the potential to be fertilized. Only matured oocytes can be fertilized. Germinal vesicles (GVs) are immature oocytes that do not have the ability to be fertilized. Atretic oocytes or oocytes with broken zonas are not viable and are unable to be fertilized [8, 9].

Under normal standard situation, the sperm after capacitation must have the ability to reach a mature oocyte (egg) and enter the oocyte, activating matured MII oocyte to form pronucleus. At the same time, the oocyte must have ability to block more than one sperm to enter oocyte to avoid to form polyspermy [10]. If sperm does not have ability to move or penetrate oocyte zona pellucida, the intracytoplasmic sperm injection (ICSI) technique can overcome this obstacle.

If it is normal fertilization, one prenucleus (PN) is from the oocyte, and another is from the sperm. The oocyte and sperm are haploid, in which each contains only ½ of the normal genetic material of a human, and their combination recover their adult body chromosome numbers. During this fertilization process, one important condition is that oocyte must be matured MII, and sperm must have ability to enter eggs. Also, sperm must activate oocytes to form pronuclei. When the sperm enters the oocyte, its head must de-condense, allowing the chromosomes to unwind. The membrane surrounding the sperm must be broken for the sperm head to de-condense [11]. This is the normal traditional IVF process when the sperm tail detaches from the head. However, ICSI needs to bend the sperm tail using the ICSI pipet by breaking the membrane surrounding the sperm. Thus, normal fertilized oocytes or zygotes should contain two prenuclie (2PN) with two polar bodies at 16–18 hours post-insemination. Time-lapse monitor has revealed 17 ± 1.0 hour after insemnination as the best observation for fertilization [12]. However, routine human IVF laboratories always have many abnormal fertilization eggs including zero PN (0PN) unfertilization, 1PN, 3PN, and multi-PN phenomena (**Figure 1**). Abnormal fertilization may be due to a number of reasons. For example, a 1PN could be caused by the failure of the sperm to de-condense or by failure of the oocyte to activate. 3PNs can be caused by the failure of the oocyte to block more than one sperm from entering the oocyte or by the failure of the oocyte to extrude the second polar body [13, 14].

Because these unfertilized oocytes and abnormal fertilized oocytes have no potential for producing a viable pregnancy, they are immediately discarded in most human IVF laboratories. However, recent many studies have found that some abnormal pronuclear-stage embryos are not necessarily genetically defective and have potential development capability to become blastocysts [7, 15]. Furthermore, normal pregnancies derived from monopronulcear (1PN) zygotes have been reported [4, 16, 17]. Abnormal fertilization oocytes may produce healthy babies [4]. Thus, the exploring

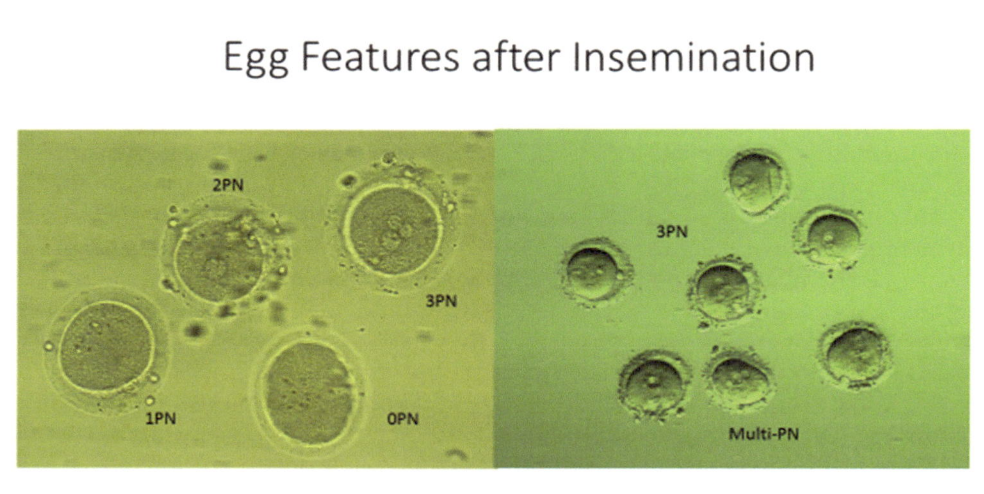

Figure 1.
Human oocytes show several different features after insemination: Unfertilized egg (0PN), normal fertilization (2PN), abnormal fertilization (3PN, multi-PN).

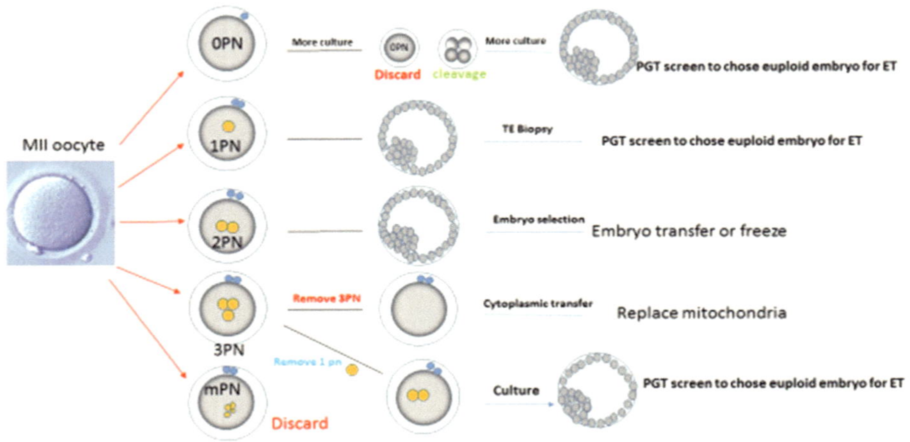

Figure 2.
After fertilization, the tendency of oocyte development shows no fertilization (0PN), Normal fertilization (2PN), and abnormal fertilization (3PN, multi-PN). 0PN and 1PN oocytes may be further cultured to cleavage stage and blastocyst stage. Normal 2PN embryos may be used for transfer or cryopreservation. 3PN zygotes may be enucleated for cytoplasmic transfer or one pronucleus may be enucleated from 3PN eggs to form normal 2PN embryos. Multi-PN oocytes should be discarded immediately. Abnormal fertilization embryos should be diagnosed by PGT to select normal chromosome composition embryos for transfer.

application of these abnormal eggs will be significant for rare egg patients. We have drawn an outline for the development and application of these abnormal fertilization oocytes (**Figure 2**).

3. No pronuclear (0PN) oocytes

At fertilization check, the embryologist usually observed that some of the eggs have no fertilization, which shows 0PN. This often presents two situations. One is immature oocyte without a polar body or staying germinal vesicle stage, which showed that these immature oocytes will not be capable of fertilization at a later time. Further, those oocytes usually do not have potential development capability on the next day. However, those MI oocytes without polar body at the oocyte retrieval day may be cultured for an extended time to observe their polar body appearance. After 5–6 hour culture, the oocytes with polar body oocytes may be used for ICSI to get fertilized zygotes.

Another situation is no fertilization with mature oocytes that have not shown the characteristic formation of their pronucleus. This may have either sperm or oocyte defect in its number of chromosomes, or they do not have a complete plan for building an embryo. Some eggs in the ovaries are atretic and nonfunctional, and have no chance of developing into an embryo. Even though these oocytes are inseminated with normal sperm, they cannot show fertilization with 0PN [18], which results in the egg being incapable of further development. The sperm can also be the cause of a mature egg not fertilizing [19]. The semen sample may show a low number of sperm and abnormal morphology, and these sperm are incapable of penetrating the egg's membranes. This is often associated with a high percentage of abnormally shaped sperm (morphology) [20]. In severe cases, sperm abnormality may result in total failure of fertilization. Even though the sperm can penetrate zona pellucida to enter the egg, the oocyte still cannot form a pronucleus. Under many situations with normal sperm parameters and good oocytes, about 10–20% oocytes still display 0PN at check fertilization [16]. However, some 0PN oocytes may show parthenogenetical activation and cleave into two or more cell stage, even develop to form blastocysts during their culture [15]. A recent study with 6466 ICSI oocytes has reported that 11.2% of 0PN oocytes formed good-quality blastocysts [16]. Further, the 0PN-derived embryos in conventional IVF cycles also result in live birth [4], and about 11.3% (4966/43,949) 0PN-derived embryos and 275 0PN-derived embryos were transferred in 70 cycles and 13 healthy infants were born with a 17.0% implantation rate [16]. Other studies also respectively reported live birth rate with 0PN-derived embryos as 4.6% (13/285), 23.1% (3/13), 48.1% 913/27), 35.6% (155/435) [15–17]. These results indicate that some 0PN oocytes may develop normal embryos and transferring them may produce healthy babies. Based on current preimplantation genetic testing (PGT) application, 0PN-derived embryos may be screened to select normal embryo for transfer in the human IVF practice, especially for those patients where no other embryos are available [21].

4. Monopronuclear (1PN) eggs

During fertilization check at 16–18 hours post-insemination, about 2.7–5.5% oocytes following conventional IVF and 4.9–11.4% for ICSI procedure [6] display one pronucleus (1PN). This usually occurs when a defective sperm enters an egg, signaling the egg to form its pronucleus. Meanwhile, the defective sperm is incapable of forming its pronucleus. With only one half of the chromosomes functional, the plan for building an embryo is incomplete. These eggs are known as 1PN and have no potential for life. Embryos displaying only a single pronucleus (1PN) are widely regarded to be abnormal and not suitable for clinical use. It is therefore common practice to discard these embryos following fertilization check. These zygotes with 1PN are thought to be at a higher risk of being haploid, and the transferring these embryos may result in an implantation failure. It is not clear to identify the origin of a monopronucleated egg from either oocyte or sperm. It could be parthenogenetic oocyte activation or an abnormal formation of the nuclear envelope [22]. The latter may result either from the combination of the two genomes into a single PN or from the failure to organize a nuclear envelope around one of the parental genomes. However, 1PN embryos have shown the ability for normal blastocyst development, albeit at a decreased rate, and the ability to produce live births [23]. Recently, several articles about the possible origins of zygotes with 1PN and the chromosomal constitution of the resulting embryos have been published [21, 24]. From these reports, we may assume that a considerable number of embryos originating from 1PN zygotes could have a normal chromosomal constitution, and they could be considered for reproductive purposes in cases where no embryos deriving from normally fertilized zygotes are available. Staessen et al. [25] evaluated 312 1PN zygotes twice (at 16–18 h and at 20–24 h after insemination) and found that 25% zygotes appear the second pronucleus after 4–6 h, which showed why abnormal pronuclear-stage embryos are not necessarily genetically defective [26]. In human IVF practice, 1PN embryos may either be discarded or considered for further culture to observe if they are possibly developing to blastocyst for embryo transfer in case of absence of an adequate normally fertilized embryos (**Figure 2**).

We found that 1PN embryos with potential formed blastocysts often have normal baby birth. Under some situations, no definitive genetic evidence shows that all 1PN zygotes are chromosomally abnormal. It is possible that some may be chromosomally normal despite the morphologic defects. The 1PN embryos with abnormal chromosomal constitution are very difficult to develop to blastocyst stage. Recent preimplantation genetic testing technology may screen 1PN embryos to determine their chromosome constitution to identify haploid, diploid, or triploid. So far, many papers have reported that some 1PN embryos with normal diploid could develop to blastocysts. These euploid embryos were transferred, and healthy babies have been born [27]. Recently, Usshe et al. have reported that they transferred 175 1PN embryos and obtained 69 live births. Compared with 2PN zygotes, 1PN embryos derived by ICSI have a significantly lower blastocyst development potential and embryo utilization rate (68.3 vs. 32.2%), with a reduced ongoing pregnancy rate (27.3 vs. 28.1%) [21]. Thus, not all 1PN embryos are abnormal and some 1PN embryos are usable. However, we recommend that the 1PN zygotes should be cultured and further observed their embryo cleavage and blastocyst formation until Day 3 to Day 6. Then, these blastocysts will be diagnosed by PGT technique to choose normal chromosome diploid before embryo transfer so that normal healthy babies should be born.

5. Triploidy (3PN)

During the process of IVF insemination, a mature egg is placed with several thousand sperm together. Normally, a mature and healthy egg will allow a single sperm to penetrate oocyte zone pellucida and egg plasma membrane and to enter oocyte cytoplasm. After sperm enters egg, a chemical process immediately occurs within the egg so that oocyte can block any other sperm entering oocyte again to prevent polyspermy fertilization, even though it is surrounded by thousands of sperm. This process is known as the cortical granule reaction. There are two types of mechanisms for polyspermy block: the "zona reaction" and the "oocyte membrane block" to sperm penetration [28]. Some eggs are defective in this process, and the chemical reaction is slow or incomplete, which results in a second sperm or more sperm to enter the oocytes cytoplasm and form a third pronucleus (3PN) or multi-pronucleus (mPN) eggs. This process is known as polyspermic fertilization. Li et al. showed that more retrieved oocytes and higher HCG day peak E2 value could result in 3PN incidence more easily [29]. Abnormal fertilization leading to 3PN formation is a common phenomenon in ICSI and IVF. The 3PN zygotes are observed from time to time after conventional IVF (5 to 8.1%) [13, 30] or intracytoplasmic sperm injection (ICSI, 2.5% to 6.2%) [14, 30]. In early stages of the 3PN zygote development, cellular cleavage is likely to be normal, but their proliferation may arrest because of aneuploidy occurring at a later development stage. However, 3PN embryos are capable of self-correction to make triploid become diploid and form blastocysts [31–33]. There is a significant difference on self-correction between routine IVF and ICSI derived embryos [31]. Traditional IVF uses more than thousands of sperm, which may result in polyspermic fertilization. These embryos may contain the third set of chromosome. It is very difficult for these 3PN embryos to self-correct or eliminate a set of chromosome. However, the insemination of oocytes by direct ICSI usually eliminates two sperm to oocyte, but it does not prevent in all oocytes the formation of abnormally 3PN fertilized zygotes. These 3PN incidences are generally due to a retention of the second polar body after ICSI, or oocyte activation problem in addition to other factors that have not yet been studied in detail. Recently, Wei et al. have identified the unique genetic mechanisms underlying fertilization failure and suggested artificial oocyte activation technology with ICSI to improve fertilization rate and interfering with or supplementing the relevant genes to improve and restore infertile couple's fertility [18].

Also, those embryos derived from 3PN by ICSI have more possibility for self-correction [34]. Guar et al. showed that both ICSI-3PN and IVF-3PN embryos are capable of self-correction, but the 3PN embryos derived with ICSI have higher capability of self-correction than the those derived with IVF. The ability of a 3PN embryo to become self-corrected is determined by the parental origin of the extra pronucleus [31]. Another study reported that about 62.5% self-corrected ICSI-derived 3PN embryos could progress back to the blastocyst stage, whereas 54.5% were hetero-parental diploid blastocysts [33]. Mutia et al. observed 30 embryos with 3PN zygotes; they found that 33.3% had a normal chromosomal array, with 22 pairs of autosomes and 2 pairs of sex chromosomes. However, during other 66.7% 3PN embryos, triploidy was 43.3%, mosaicism was 13.4%, and aneuploidy was 10%. These studies show that not all morphologically 3PN embryos are genetically abnormal [35].

Based on the above analysis, although most of 3PN embryos are genetically abnormal chromosome composition, a small portion of 3PN embryos may develop to blastocyst stage and these capable development embryos may contain normal chromosome composition. Thus, some reports have showed pregnancy and subsequent delivery of a healthy newborn after the transfer of a blastocyst that developed from a tripronuclear zygotes [36].

In human IVF clinical practice, we strongly suggest that embryos derived from 3PN abnormalities be detected prior to embryo transfer. To obtain euploid embryos, preselection must be done to choose the embryos with the highest implantation potential. Except the current methods of embryo selection such as morphology, metabolomics, and morphokinetics, the most accuracy embryo screen method is preimplantation genetic testing (PGT) or preimplantation genetic screening (PGS). Then, the embryos with normal genetic composition may be selected for embryo transfer.

6. Artificial correction for 3PN embryos by removing one pronucleus

At fertilization checking, if an oocyte shows obviously three pronuclei in the cytoplasm, one pronuclues may be removed by micromanipulation technique. Some studies have attempted to restore the diploid status of 3PN zygotes by microsurgical removal of the extra pronucleus [37–40]. Some reports showed that the microsurgical removal of a pronucleus from tripronuclear human zygotes has been successful [41]. Further, Kattera et al. reported a birth of a healthy boy after transferring three embryos that developed from microsurgically corrected 3PN zygotes [42]. Our laboratory performed microsurgical removing one PN from 3PN zygotes to artificially correct this abnormal fertilization (**Figure 3**). The results showed that this procedure could improve the corrected 3PN zygote cleavage and blastocyst formation (corrected zygotes 10.5% vs. intact 3PN zygotes 6.6%), but we did not further examine the chromosome composition from these blastocyst by microsurgical correction, which needs to be further studied. Liao et al. analyzed the chromosomal constitution of microsurgically corrected 3PN embryos and indicated that after correction, the diploid rate of the blastocysts (55.0%) was significantly higher ($P < 0.05$) than that of the arrested cleavage-stage embryos with non-correction (18.4%). The triploid rate of the microsurgically corrected 3PN zygotes (5.7%) was significantly lower ($P < 0.01$) than that of intact 3PN zygotes (19.4%) [38].

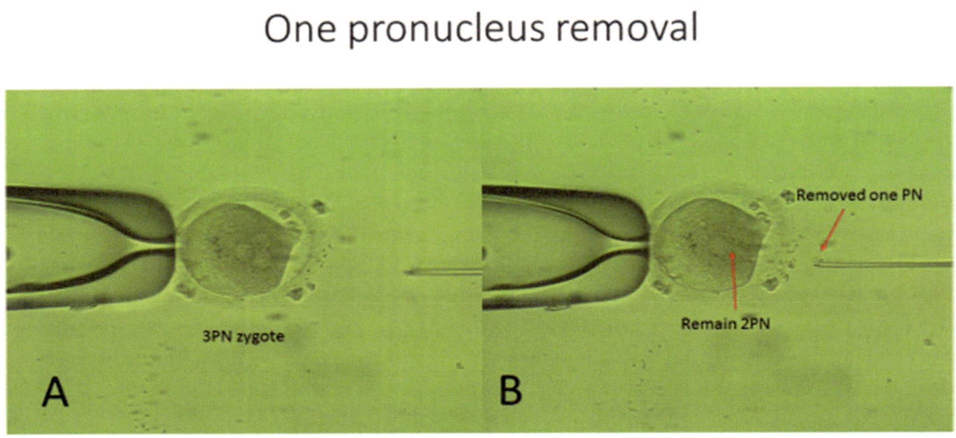

Figure 3.
Removing one pronucleus from 3PN zygote. A indicates 3PN before removing one pronucleus, and B shows 2PN after microsurgical removing one pronucleus.

Thus, we think that the microsurgical operation can effectively remove one pronucleus from 3PN zygotes. After pronuclear removal, the formed blastocyst may be screened by the current PGT technique to select diploidized blastocysts for embryo transfer in rare embryo patients. Whether these corrected embryos from 3PN zygotes can be used in the clinic practice depends on preimplantation genetic diagnosis and heteroparental identification. Transferring one normal genetic composition blastocyst with removal one pronucleus from 3 PM zygote may obtain the pregnancy similar to normal 2PN formation embryo transfer. However, this microsurgical pronuclear removal often causes some 3PN zygote damage and results in low blastocyst development. Thus, this technique is suited to just some patients without normal embryos to make pregnancy.

7. The application of 3PN cytoplasm

As described above, triploidy can appear either from more than two sperm to enter an oocyte or from retention of the second polar body extrusion because of the oocyte meiotic failure. Oocyte membrane and zona pellucida has a capability to block extra sperm entering oocyte cytoplasm. All these blocks involve a depolarization of the oocyte membrane caused by the influx of Na^+, and a Ca^{2+} oscillation event activated by the sperm attached to the oocyte membrane. Further, the increased intracellular Ca2+ concentration triggers the exocytosis of cortical granules to release their contents passing the oocyte membrane into the perivitelline space [43]. Thus, 3PN formation has a close concern with egg zona pellucida and membrane and seems not to be associated with egg cytoplasm. This phenomena prompts us to consider the similarity of ooplasm between 3PN and 2PN zygotes [30]. Recently, Kotil et al. have done a detail research on ooplasm difference between 3PN and 2PN zygotes, and they examined actin filament and γ-tubulin localization by immunohistochemistry, evaluated the inner membrane potential of mitochondria by JC-1 staining, and observed oocyte ultrastructure by transmission electron microscopy [44]. Their results showed too much similarity of all ooplasms, especially the mitochondrial membrane potential between 3PN and 2PN zygotes. This finding provides us an idea how to use 3PN zygote ooplasm to correct some advance woman eggs with dysfunction of mitochondria.

In human IVF practice, many aging women still are interested in having their babies from assisted reproductive technology. However, these women often show diminished ovarian reserve symptom, which is associated with reduced quantity and quality of the retrieved oocytes and failure of IVF outcome. For over-40-year-old women, their oocyte quality has a significant reduction. A major reason is due to the lack of enough ooplasm component parts in oocyte cytoplasm, such as synthesis of proteins (microtubules), which supports cytoskeleton organization and mitochondria relocation. The functional competency of organelles is crucial for the quality of oocyte. The mitochondria are responsible for oocyte energy production. Thus, aging women often have low quality of oocyte because of their mitochondria malfunction, which results in aging women's decreased IVF success rate [44].

In order foraged women to realize their dream to have babies, the only approach is to use donation of eggs, but most aging women do not like to seek for oocyte donation. Thus, we must search for new ways to overcome the problem of aging oocytes in our IVF laboratory. It has been reported that along aging with the decline in oocyte quality, oocyte mitochondrial DNA (mtDNA) shows an obvious defects [45]. The repair of oocyte mtDNA may improve oocyte quality. To date, IVF physicians often give patients the supplement of various compounds, such as antioxidants and coenzymes to improve mitochondrial function and fertility outcome. In the last few years, nuclear transfer has become another alternative treatment approach to improve age women oocyte quality and this tri-parent baby has been born [46]. Germline nuclear transfer is a novel technology on ART that involves the transfer of nuclear genome from an affected oocyte/zygote of the patient to the cytoplast of an enucleated donor oocytes/zygote [47]. Current developed nuclear transfer techniques contain mitochondrial transfer, pronuclear transfer, spindle transfer, blastomere transfer, ooplasmic or cytoplasmic transfer, polar body transfer, and so forth [48, 49].

Nuclear transfer may be performed in either germinal vesicle (GV) stage oocytes [50] or metaphase II stage oocytes [49]. Mature oocytes may be used for micro-spindle transfer [51]. Fertilized eggs may be used for two pronuclear transfer or cytoplasmic replacement. However, all these nuclear transfer procedures need to be performed in third younger partner donor eggs. Using donor eggs is expensive and needs more time based on donor agreement consent. Based on the similarity of ooplasm between 3PN and 2PN zygotes, we may use discarded 3PN zygotes to provide good cytoplasm to replace advance aging women oocyte cytoplasm and help to bring about aging women's embryo growth.

Ooplasmic transfer is one of the latest innovations in IVF treatment, which involves transferring a small amount of oocyte cytoplasm from normal fertile women into oocytes of women with damaged mitochondria. It has been reported that the children born using this technique will possess cytoplasmic organelles including mitochondria from both their biological mother and donor ooplasm [52]. This is called tri-parents baby who has the mixing of parent and cytoplasmic donor mitochondria. However, it is not easy to obtain donor oocytes for this ameliorated purpose. Recently, Fujimine-Sato et al. have tried to use the cytoplasmic function of abnormally fertilized embryos by pronuclear-stage cytoplasmic transfer to improve oocyte quality [53]. In our clinic practice, we may take out partial cytoplasm of the discard 3PN and inject to the aged woman oocyte to form two kinds of mixed cytoplasm in an embryo (**Figure 4**). Our primary experience shows that this procedure may improve aging embryo quality. However, this part of ooplasmic transfer does not change aging women's genetic composition, or inheritable genetic modification. Although an aging woman's modified embryos still keep her with her husband's genetic genes in the cell

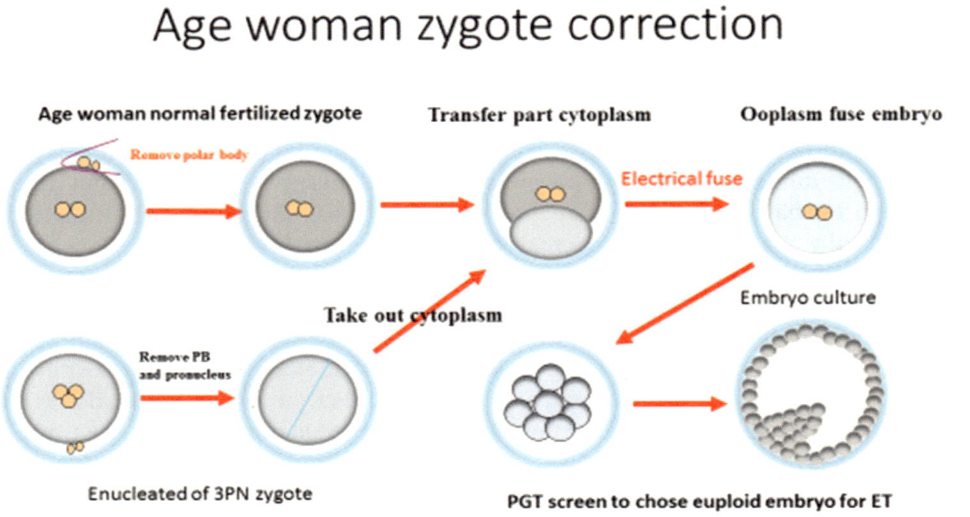

Figure 4.
Advance aging woman oocyte correction. After fertilization, firstly two polar bodies have been removed from normal fertilized zygote. Then, all 3 pronuclei are removed from 3PN zygote and about 30–50% cytoplasm from enucleated 3PN zygote will be taken out and transferred this partial ooplasm into aging woman zygote. After two zygote cytoplasm fusing by electrical purse, one mix embryo of cytoplasm will be formed for further culture. After PGT selection, a normal embryo may be transferred.

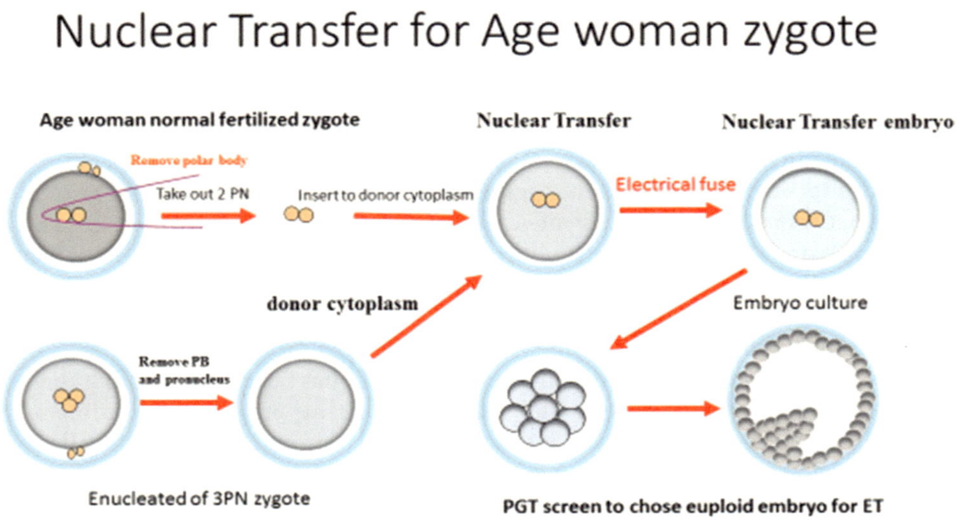

Figure 5.
2PN nuclear transfer using 3PN zygote cytoplasm. Because the aged woman 2PN embryo has dysfunction of ooplasm, especially their mitochondria malfunction, their 2PN may be transferred to enucleated 3PN cytoplasm so that using 3PN ooplasm may improve aged women's 2PN normal development.

nucleus, but the baby born with this technique might appear epigenetics because the mixed mitochondrial DNA would be passed on to all future generations.

Also, 3PN zygote may provide its cytoplasm for nuclear transfer. Yao et al. performed a study about ooplast transfer of triploid pronucleus zygote to reconstruct human-goat embryonic development in which their results showed the 3PN cytoplasmic transfer could significantly improve the early development of humanized new constructed embryos [54]. In human IVF laboratory, we may use discarded 3PN zygote cytoplasm to replace advance aging woman ooplasm to create a new zygote by transfer age woman 2PN into enucleated 3PN cytoplasm (**Figure 5**). This technology does not use a woman's coplasm. Although it do not change the woman's original 2PN genetic constitution, maternal cytoplasm is from a donor. Thus, the donor 3PN cytoplasm must have the effect of epigenetics in this offspring. This technology and outcome on human IVF practice application still need more research in the future.

8. Conclusion

Human IVF laboratory often shows many abnormal fertilization eggs, such as 0PN, 1PN, 3PN, and multi-PN zygotes, and these abnormal eggs are usually not used and typically discarded. Recent research and our observation have showed that further work on these abnormal eggs may provide some hope for advance age infertile couples to realize their desire to have babies in their family. We recommend that 0PN eggs may be done for rescue ICSI or further cultured to next few days to observe their further cleavage and development to blastocyst. Selecting chromosome normal blastocyst for embryo transfer may have healthy baby birth. Many studies showed that not all 1PN embryos are abnormal and some 1PN embryos are usable. The 1PN zygotes should be cultured and further observed their embryo cleavage and blastocyst formation until Day 3 to Day 6. Then, these blastocysts will be diagnosed by PGT technique to choose normal chromosome diploid before embryo transfer so that normal healthy babies should be born. Although most of 3PN embryos are genetically abnormal chromosome composition, a small portion of 3PN embryos may develop to blastocyst stage and these capable development embryos may contain normal chromosome composition. Also, those embryos derived from 3PN, especially by ICSI, have more possibility for self-correction to become normal euploid embryos. The microsurgical removing one PN from 3PN zygotes may artificially correct this abnormal fertilization. After one pronuclear removal, the formed blastocyst may be screened by current PGT technique to select diploidized blastocysts for embryo transfer in rare embryo patients. Some reports have showed pregnancy and subsequent delivery of a healthy newborn after the transfer of the corrected 3PN blastocyst. Because of no obvious difference of ooplasm between normal fertilized 2PN zygotes and 3PN zygotes, this cytoplasm of 3PN zygotes may be used to improve advance age woman poor oocytes and embryo quality. Transferring partial cytoplasm from 3PN to the fertilized 2PN zygote of aged woman may promote her embryo to develop blastocysts. This partial ooplasmic transfer does not change an aging woman's genetic composition; thus, the aging woman's embryos still keep her with her husband's genetic genes in the cell nucleus. However, the baby born with this technique might appear epigenetics because the mixed mitochondrial DNA would be passed on to all future generations.

Author details

Bin Wu[1*], Xue Feng[2], Suzhen Lu[1] and Timothy J. Gelety[1]

1 Arizona Center for Reproductive Endocrinology and Infertility, Tucson, Arizona, USA

2 The Assisted Reproduction Center, Northwest Women's and Children's Hospital, Affiliated with Xi'an Jiaotong University, Xi'an, Shaanxi, China

*Address all correspondence to: bwu13@yahoo.com

References

[1] Kemper JM, Liu Y, Afnan M, Mol BWJ, Morbeck. What happens to abnormally fertilized embryos? A scoping review. Reproductive Biomedicine Online. 2023;**46**(5):802-807. DOI: 10.1016/j.rbmo.2023.02.005

[2] Reichman DE, Jackson KV, Racowsky C. Incidence and development of zygotes exhibiting abnormal pronuclear disposition after identification of two pronuclei at the fertilization check. Fertility and Sterility. 2010;**94**:965-969

[3] Xiao C, Shi S, Mao J, Zou L, Keda Y. Developmental potential of abnormally fertilized oocytes and the associated clinical outcomes. Frontiers in Physiology. 2020;**11**:528424. DOI: 10.3389/fphys.2020.528424

[4] Capalbo A, Treff N, Cimadomo D, Scott R, Ubaldi FM, Rienzi L. Abnormally fertilized oocytes can result in healthy live births: Improved genetic technologies for preimplantation genetic testing can be used to rescue viable embryos in in vitro fertilization cycles. Fertility and Sterility. 2017;**108**(6): 1007-1015. DOI: 10.1016/j.fertnstert.2017.08.004

[5] Destouni A, Dimitriadou E, Masset H, Debrock S, Melotte C, Bogaert KVD, et al. Genome-wide haplotyping embryos developing from 0PN and 1PN zygotes increases transferrable embryos in PGT-M. Human Reproduction. 2018;**33**:2302-2311

[6] Feenan K, Herbert M. Can 'abnormally' fertilized zygotes give rise to viable embryos? Human Fertility (Cambridge, England). 2006;**9**:157-169

[7] Yao G, Xu J, Xin Z, et al. Developmental potential of clinically discarded human embryos and associated chromosomal analysis. Scientific Reports. 2016;**6**:23995. DOI: 10.1038/srep23995

[8] Jamnongjit M, Hammes SR. Oocyte maturation: The coming of age of a germ cell. Seminars in Reproductive Medicine. 2005;**23**(3):234-241. DOI: 10.1055/s-2005-872451

[9] Verlhac MH, Terret ME. Oocyte maturation and development. F1000Res. 2016;**5**:F1000 Faculty Rev-309. DOI: 10.12688/f1000research.7892.1

[10] Wang WH, Day BN, Wu GM. How does polyspermy happen in mammalian oocytes? Microscopy Research and Technique. 2003;**61**(4):335-341. DOI: 10.1002/jemt.10346

[11] Ikawa M, Inoue N, Benham AM, Okabe M. Fertilization: A sperm's journey to and interaction with the oocyte. The Journal of Clinical Investigation. 2010;**120**(4):984-994. DOI: 10.1172/JCI41585

[12] Lee MJ. Time of embryo cleavage. In: Wu B, editor. Embryo Cleavage. BoD – Books on Demand. Rijeka, Croatia: InTech; 2017. pp. 17-23

[13] Porter R, Han T, Tucker MJ, Graham J, Liebermann J, Sills ES. Estimation of second polar body retention rate after conventional insemination and intracytoplasmic sperm injection: in vitro observations from more than 5000 human oocytes. Journal of Assisted Reproduction and Genetics. 2003;**20**:371-376

[14] Rosenbusch BE. Mechanisms giving rise to triploid zygotes during assisted

reproduction. Fertility and Sterility. 2008;**90**:49-55

[15] Paz MV, Chiera M, Hovanyecz P, Cicaré J, Perfumo P, Domenech L, et al. Blastocysts derived from 0PN oocytes: Genetic and clinical results. JBRA Assisted Reproduction. 2020;**24**:143-146

[16] Liu J, Wang XL, Zhang X, Shen CY, Zhang Z. Live births resulting from 0PN-derived embryos in conventional IVF cycles. Journal of Assisted Reproduction and Genetics. 2016;**33**(3):373-378. DOI: 10.1007/s10815-015-0644-6

[17] Kobayashi T, Ishikawa H, Ishii K, et al. Time-lapse monitoring of fertilized human oocytes focused on the incidence of 0PN embryos in conventional in vitro fertilization cycles. Scientific Reports. 2011;**11**:18862. DOI: 10.1038/s41598-021-98312-1

[18] Wei Y, Wang J, Qu R, Zhang W, Tan Y, Sha Y, et al. Genetic mechanisms of fertilization failure and early embryonic arrest: A comprehensive review. Human Reproduction Update. 2024;**30**(1):48-80. DOI: 10.1093/humupd/dmad026

[19] Safari H, Anbari F, Ghasemi-Esmailabad S, Maleki B, Dehghan Marvast L, Reza TA. Relationship between sperm quality and total fertilization failure in intracytoplasmic sperm injection and in vitro fertilization cycles: A cross-sectional study. International Journal of Reproductive BioMedicine. 2022;**20**(5):413-422. DOI: 10.18502/ijrm.v20i5.11056

[20] Pantos K, Sfakianoudis K, Maziotis E, Rapani A, Karantzali E, Gounari-Papaioannou A, et al. Abnormal fertilization in ICSI and its association with abnormal semen parameters: A retrospective observational study on 1855 cases. Asian Journal of Andrology. 2021;**23**(4):376-385. Published online 2021. DOI: 10.4103/aja.aja_84_20

[21] Ussher C, Holden S, Bowman M, McArthur S. What is the true potential of 1PN embryos? A report on utilisation and live birth rates following PGT-A and biparental testing. Human Reproduction. 2023;**38**(Suppl.1):O-180. DOI: 10.1093/humrep/dead093.221

[22] Bos-Mikich A, Bressan FF, Ruggeri RR, Watanabe Y, Meirelles FV. Parthenogenesis and human assisted reproduction. Stem Cells International. 2016;**2016**:1970843. DOI: 10.1155/2016/1970843

[23] Itoi F, Asano Y, Shimizu M, Honnma H, Murata Y. Birth of nine normal healthy babies following transfer of blastocysts derived from human single-pronucleate zygotes. Journal of Assisted Reproduction and Genetics. 2015;**32**(9):1401-1407. DOI: 10.1007/s10815-015-0518-y

[24] Kai Y, Iwata K, Iba Y, Mio Y. Diagnosis of abnormal human fertilization status based on pronuclear origin and/or centrosome number. Journal of Assisted Reproduction and Genetics. 2015;**32**(11):1589-1595. DOI: 10.1007/s10815-015-0568-1

[25] Staessen C, Van Steirteghem AC. The chromosomal constitution of embryos developing from abnormally fertilized oocytes after intracytoplasmic sperm injection and conventional in-vitro fertilization. Human Reproduction. 1997;**12**(2):321-327

[26] Bradley CK, Traversa MV, Hobson N, Gee AJ, McArthur SJ. Clinical use of monopronucleated zygotes following blastocyst culture and preimplantation genetic screening, including verification of biparental chromosome inheritance.

Reproductive BioMedicine Online. 2017;**34**(6):567-574

[27] Li M, Huang J, Zhuang X, Lin S, Dang Y, Wang Y, et al. Obstetric and neonatal outcomes after the transfer of vitrified-warmed blastocysts developing from nonpronuclear and monopronuclear zygotes: A retrospective cohort study. Fertility and Sterility. 2021;**115**:110-117

[28] Cheeseman L, Boulanger J, Bond L, et al. Two pathways regulate cortical granule translocation to prevent polyspermy in mouse oocytes. Nature Communications. 2016;**7**:13726. DOI: 10.1038/ncomms13726

[29] Li M, Zhao W, Xue X, Zhang S, Shi W, Shi J. Three pro-nuclei (3PN) incidence factors and clinical outcomes: A retrospective study from the fresh embryo transfer of in vitro fertilization with donor sperm (IVF-D). International Journal of Clinical and Experimental Medicine. 2015;**8**(8):13997-14003

[30] Pieters MH, Dumoulin JC, Ignoul-Vanvuchelen RC, Bras M, Evers JL, Geraedts JP. Triploidy after in vitro fertilization: Cytogenetic analysis of human zygotes and embryos. Journal of Assisted Reproduction and Genetics. 1992;**9**:68-76

[31] Grau N, Escrich L, Martin J, Rubio C, Pellicer A, Escriba M. Self-correction in tripronucleated human embryos. Fertility and Sterility. 2011;**96**(4):951-956

[32] Orvieto R, Shimon C, Rienstein S, et al. Do human embryos have the ability of self-correction? Reproductive Biology and Endocrinology. 2020;**18**:98. DOI: 10.1186/s12958-020-00650-8

[33] Grua N, Escrich L, Galiana Y, Mesequer M, garc S, Escrib M. Morphokinetics as a predictor of self-correction to diploidy in tripronucleated intracytoplasmic sperm injection-derived human embryos. Fertility and Sterility. 2015;**104**(3):728-735

[34] Berger DS, Hoff HS. Fertilization assessment in IVF and ICSI. In: Montag MHM, Morbeck DE, editors. Principles of IVF Laboratory Practice: Laboratory Set-up, Training and Daily Operation. London, UK: Cambridge University Press; 2023. pp. 207-212

[35] Mutia K, Wiweko B, Iffanolida PA, Febri RR, Muna N, Riayati O, et al. The frequency of chromosomal Euploidy among 3PN embryos. Journal of Reproduction & Infertility. 2019;**20**(3):127-131

[36] Yalcınkaya E, Özay A, Ergin E, Öztel Z, Ozornek H. Live birth after transfer of a tripronuclear embryo: An intracytoplasmic sperm injection as a combination of microarray and time-lapse technology. Journal of Turkish Society of Obstetric and Gynecology. 2016;**13**(2):95-98

[37] Rawlins RG, Binor Z, Radwanska E, Dmowski WP. Microsurgical enucleation of tripronuclear human zygotes. Fertility and Sterility. 1988;**50**:266-272

[38] Liao HQ, OuYang Q, Zhang SP, Cheng DH, Lu GX, Lin G. Pronuclear removal of tripronuclear zygotes can establish heteroparental normal karyotypic human embryonic stem cells. Journal of Assisted Reproduction and Genetics. 2016;**33**(2):255-263. DOI: 10.1007/s10815-015-0634-8

[39] Jin HX, Dai SJ, Song WY, Yao GD, Shi SL, Sun YP. Embryo developmental potential of microsurgically corrected human three-pronuclear zygotes. Systems Biology in Reproductive Medicine. 2015;**61**(2):96-102. DOI: 10.3109/19396368.2014.986693

[40] Gordon JW, Grunfeld L, Garrisi GJ, Navot D, Laufer N. Successful microsurgical removal of a pronucleus from tripronuclear human zygotes. Fertility and Sterility. 1989;**52**:367-372

[41] Malter HE, Cohen J. Embryonic development after microsurgical repair of polyspermic human zygotes. Fertility and Sterility. 1989;**52**:373-380

[42] Kattera S, Chen C. Normal birth after microsurgical enucleation of tripronuclear human zygotes: Case report. Human Reproduction. 2003;**18**:1319-1322. DOI: 10.1093/humrep/deg262

[43] Wakai T, Vanderheyden V, Fissore RA. Ca^{2+} signaling during mammalian fertilization: Requirements, players, and adaptations. Cold Spring Harbor Perspectives in Biology. 2011;**3**(4):a006767. DOI: 10.1101/cshperspect.a006767

[44] Kotil T, Kervancıoğlu ME, Kanten GE, Tunalı G, Solakoğlu S. Mitochondrial activity and cytoskeleton organization in three pronuclei oocytes after intracytoplasmic sperm injection. Zygote. 2018;**26**(4):319-325. DOI: 10.1017/S0967199418000278

[45] van der Reest J, Cecchino GN, Haigis MC, Kordowitzki P. Mitochondria: Their relevance during oocyte ageing. Ageing Research Reviews. 2021;**70**:1568-1637

[46] Nature News. First UK Children Born Using Three-Person IVF: What Scientists Want to Know First UK Baby with DNA from Three People Born after New IVF Procedure. London, UK: Nature News; 2023. Available from: https://www.nature.com/articles/d41586-023-01585-x

[47] Christodoulaki A, Boel A, Tang M, De Roo C, Stoop D, Heindryckx B. Prospects of germline nuclear transfer in women with diminished ovarian reserve. Frontiers in Endocrinology. 2021;**12**:635370. DOI: 10.3389/fendo.2021.635370

[48] Morimoto Y, Gamage USK, Yamochi T, Saeki N, Morimoto N, Yamanaka M, et al. Mitochondrial transfer into human oocytes improved embryo quality and clinical outcomes in recurrent pregnancy failure cases. International Journal of Molecular Sciences. 2023;**24**(3):2738. DOI: 10.3390/ijms24032738

[49] Mobarak H, Heidarpour M, Tsai PJ, Rezabakhsh A, Rahbarghazi R, Nouri M, et al. Autologous mitochondrial microinjection; a strategy to improve the oocyte quality and subsequent reproductive outcome during aging. Cell & Bioscience. 2019;**9**:95. DOI: 10.1186/s13578-019-0360-5

[50] Takeuchi T, Ergün B, Huang TH, Rosenwaks Z, Palermo GD. A reliable technique of nuclear transplantation for immature mammalian oocytes. Human Reproduction. 1999;**14**(5):1312-1317. DOI: 10.1093/humrep/14.5.1312

[51] Costa-Borges N, Spath K, Miguel-Escalada I, Mestres E, Balmaseda R, Serafín A, et al. Maternal spindle transfer overcomes embryo developmental arrest caused by ooplasmic defects in mice. eLife. 2020;**9**:e48591. DOI: 10.7554/eLife.48591

[52] Tanaka A, Watanabe S. Can cytoplasmic donation rescue aged oocytes? Reproductive Medicine and Biology. 2018;**18**(2):128-139. DOI: 10.1002/rmb2.12252

[53] Fujimine-Sato A, Kuno T, Higashi K, Sugawara A, Hiraga H, Takahashi A, et al. Exploration of the cytoplasmic function of abnormally fertilized

embryos via novel pronuclear-stage cytoplasmic transfer. International Journal of Molecular Sciences. 2021;**22**:8765. DOI: 10.3390/ijms22168765

[54] Yao L, Wang P, Liu J, Chen J, Hailiang Tang H, Sha H. Ooplast transfer of triploid pronucleus zygote improve reconstructed human-goat embryonic development. International Journal of Clinical and Experimental Medicine. 2014;7(10):3678-3686, Available from: www.ijcem.com, ISSN:1940-5901/ IJCEM0001730

Chapter 3

Noninvasive Testing of Preimplantation Embryos in Assisted Reproductive Technology

Qing Zhou and Yutong Wang

Abstract

One approach to improving the success of assisted reproductive technology (ART) is the careful selection of embryos prior to implantation. Although preimplantation genetic testing (PGT) is widely employed for embryo selection, it needs embryo biopsy and is detrimental to embryos. Thus, noninvasive testing of preimplantation embryos offers new possibilities for evaluating embryo quality. Here, we reviewed current progression of noninvasive embryo testing technologies, including the use of microscopy images combined with artificial intelligence (AI) to select embryos based on morphology, minimally invasive and noninvasive PGT of blastocoel fluid and spent embryo culture medium, and omics analysis of molecules in the culture medium to assess the developmental potential of embryos. More importantly, using the AI technology based on various type of data of each embryo will greatly improve the noninvasive embryo assessments. Thus, these cutting-edge technologies offer fresh insights into noninvasive testing of preimplantation embryos and have the potential to enhance the quality and efficiency of ART procedures.

Keywords: assisted reproductive technology, artificial intelligence, spent culture medium, noninvasive testing, embryo quality

1. Introduction

Infertility is a critical aspect of reproductive health and remains a highly prevalent global condition, affecting approximately 12.6–17.5% reproductive-aged couples worldwide [1]. In many cases, couples with infertility may opt for assisted reproductive technology (ART) to increase their chances of conception. ART encompasses various procedures, such as the retrieval of gametes from both the woman and man, the fertilization of multiple oocytes through *in vitro* fertilization (IVF) or intracytoplasmic sperm injection (ICSI), the culture of embryos, and the transfer of embryos back to the woman's uterus for implantation [2]. The growing use of ART treatments worldwide has sparked scientific and public interest in its effectiveness. The low success rate of ART can be partially attributed to the assessment of preimplantation embryos for selection [3, 4], as the majority of transferred embryos fail to implant in the uterus. Therefore, the development of noninvasive testing methods for embryo assessment would greatly enhance ART treatments.

2. Microscopy images coupled with artificial intelligence for embryo selection based on morphology

Under the current practice, ART treatment requires the fertilization of multiple oocytes. One of the main challenges is to select suitable embryos for transfer. Time-lapse (TL) imaging is a powerful tool for assessing embryos. The TL system monitors embryos in incubators with a built-in microscope and captures images every 5–20 minutes. These images are then assembled into a video, which is used for morphologic evaluation [5]. Typically, embryo selection involves a visual analysis by embryologists, who discard poor-quality and slow-growing embryos. A grading system for assessing blastocyst morphology quality is considered. The main morphologic criteria used are those described by Gardner and Sakkas [6], which are evaluated using three parameters: the developmental stage of embryos, the quality of the inner cell mass (ICM), and the quality of the trophectoderm (TE). However, embryo grading relies on the analysis of embryologists and is known to be subjective and susceptible to human error due to significant inter- and intra-observer variability [7].

In recent years, artificial intelligence (AI) has been extensively utilized to enhance and automate the procedure of ranking embryos. A convolutional neural network (CNN) is a type of deep learning algorithm that excels in processing 2-dimensional image data and analyzing extensive datasets in a grid pattern [8]. This neural network is commonly employed for medical imaging tasks and embryo assessment. By extracting pertinent information from embryo microscopy images and monitoring videos, AI has the potential to decrease the variability among embryologists and offer a precise and dependable interpretation of embryo morphology, as well as potentially predicting embryo viability [9, 10].

However, it is still crucial to assess the performance of these AI systems in comparison to the evaluations made by embryologists when selecting embryos. By employing a deep neural network (DNN)-based AI approach, Khosravi et al. [11] developed a framework (STORK) to predict blastocyst quality using more than 50,000 human embryo images, with an area under the receiver operating characteristic (ROC) curve (AUC) greater than 0.98. Additionally, they invited five embryologists from three different clinics to evaluate over 300 embryos from various laboratories. As anticipated, there was a low level of agreement among the embryologists, while the STORK accurately predicted the majority vote of the embryologists (at least three out of the five) with a precision of 95.7%. Their findings suggest that deep learning approaches can provide precise quality assessments of blastocysts in different clinical conditions. Similarly, another AI model for ranking blastocyst embryos using deep learning, trained on a large and diverse dataset, demonstrated the potential for improved prediction of clinical pregnancy [12]. A recent review article has encapsulated around 20 research papers on the use of AI in grading embryos. It has been concluded that the AI model achieved a median accuracy of 75.5% (range 59–94%) in predicting the grade of embryo morphology [10]. Furthermore, when clinical information inputs were incorporated, the AI models exhibited a higher median accuracy of 81.5% (range 67–98%).

The current findings indicate that AI models can outperform embryologists in terms of assessing embryo morphology, and using clinical data could enhance the performance of these models. Future studies could explore other potential applications of AI models, such as evaluating the ploidy status of embryos and predicting their implantation potential, which are the most relevant clinical outcomes in ART.

3. Noninvasive preimplantation genetic testing of embryos

3.1 Preimplantation genetic testing of embryos

As mentioned above, morphological assessment is the most commonly used method for selecting embryos, and various AI-based grading systems have been developed to enhance the evaluation of human blastocysts. However, it is worth noting that nearly half of embryos with good morphology are aneuploid [13], indicating that relying solely on morphological grading is inadequate for embryo selection, especially in terms of chromosomal assessments. Embryonic aneuploidy is a significant factor that impacts the success rate of assisted pregnancies and is responsible for over 50% of abortions [14]. Clinically, these embryonic chromosomal abnormalities can be prevented by conducting preimplantation genetic testing for aneuploidy (PGT-A). To carry out this test, embryonic biopsies are typically obtained at the cleavage stage (Day 3) or the blastocyst stage (Day 5–7). Subsequently, array-based comparative genomic hybridization (aCGH) or next-generation sequencing (NGS) platform can be utilized to identify numerical chromosome copy number aberrations. Clinical efficacies of PGT-A include increased rates of implantation and decreased risk of clinical miscarriage, particularly for patients with advanced maternal age (AMA) and recurrent implantation failure (RIF) and couples who have experienced recurrent pregnancy loss (RPL) [15, 16].

In addition to the traditional invasive PGT-A using cells from the TE, the technology of this test has evolved to include minimally invasive PGT-A (miPGT-A) using fluid from the blastocoel cavity and noninvasive PGT-A (niPGT-A) based on the spent culture medium (SCM) of embryos [17].

3.2 miPGT-A of blastocoel fluid

Blastocoel fluid (BF) is a fluid that accumulates in the blastocoel cavity. In 2013, Palini et al. [18] first reported the observation of DNA in the BF of embryos. Fragments of DNA shed from the embryo into this fluid result in a potential material resource for genetic analysis of PGT-A. Gathering BF involves inserting a needle between cell junctions into the blastocoel cavity and aspirating the fluid, instead of performing an embryo biopsy. Although it is still considered invasive, it appears to be well tolerated by the embryo [19]. However, the concordance rate between BF and TE biopsy has varied, ranging from no agreement with TE PGT-A to a significant correlation as high as 97.4% [20, 21]. There are various explanations for this wide range of concordance values, such as the unclear source of detected DNA fragments and different protocols used in multiple embryology laboratories. While there is a hypothesis that DNA in the BF may better represent the embryos compared to what is secreted into the culture medium [18], the concordance results differ from those of the SCM. Additionally, the SCM is collected after the embryo has been removed, making it truly noninvasive.

3.3 niPGT-A of spent culture medium

The culture medium is the liquid in which the preimplantation embryo will grow during its crucial early stages of development. In 2013, Stigliani et al. [22] made the initial discovery that human preimplantation embryos release genomic and mitochondrial DNA into the culture medium during *in vitro* culture. They also confirmed that the contents of the DNA were linked to embryo fragmentation. Since then, the potential clinical use of these cell-free DNA (cfDNA) has become a matter of great concern in the field of ART, especially in the development of noninvasive technology for determining chromosomal abnormalities [21, 23, 24]. Although the presence of cfDNA in SCM has been confirmed, achieving chromosomal screening would require high sensitivity and reproducibility of whole-genome amplification (WGA) for these low-amount DNA in SCM. Xu et al. [25] reported a noninvasive chromosomal screening (NICS) method by sequencing the genomic DNA in SCM of human blastocysts, using multiple annealing- and looping-based amplification cycles (MALBAC). The results of the NICS assay were validated by comparing each result with the chromosome information directly obtained from the corresponding embryos. The NICS assay has a high correlation with the identification of aneuploidy, making it potentially suitable for a wider application of PGT-A due to its noninvasiveness.

The main challenge of developing a niPGT-A technique was to verify the consistency between samples and the traditional embryo TE biopsy. Several recent studies in randomized clinical trials have primarily investigated the clinical effectiveness of niPGT-A. Some studies have highlighted the challenges associated with this technique, while others have demonstrated its potential for the future.

In a clinical study, BF samples exhibited high rates of amplification failure, resulting in an overall concordance rate as low as 37.5%. While samples from SCM performed better than BF samples, genetic analysis of SCM samples showed a high detection of artifacts due to the risk of maternal contamination. Therefore, SCM should not be utilized for the detection of single-gene mutations [26]. Another study also reported a high rate of DNA amplification failure in niPGT-A samples. In addition, there were 42 cases (40.4%) where the whole-chromosome discordance with TE biopsy was observed, rendering the clinical applicability of niPGT-A virtually impossible [27]. With further improvements, DNA can now be detected in the majority of SCM samples. This allows for the quantitative and qualitative measurement of segmental aneuploidy or mosaicism caused by individual chromosomal aberrations. However, TE biopsy still remains a more accurate and reliable methodology for PGT-A [28]. Although cfDNA in SCM shows potential as a safe and simple strategy for PGT-A, its clinical application for diagnostic purposes still requires further validation.

To promote the clinical application of niPGT-A, various validation studies have been conducted and have reported the comparison results of niPGT-A with TE biopsy. Jiao et al. [29] reported that niPGT-A can provide fast and accurate results, with the overall testing time reduced to 7.5 hours and a concordance of approximately 90% with results from embryo samples. These advanced findings not only indicate the effectiveness of niPGT-A but also suggest its potential for achieving fresh blastocyst transfer in the future. In addition to higher concordance rates for embryo ploidy and chromosome copy numbers, niPGT-A also demonstrates higher positive predictive value (PPV) and specificity for embryonic mosaicism compared to TE biopsy [19]. Consistent results were also observed in a large-scale validation study of approximately 260 embryos, promising niPGT-A as a reliable assay for prioritizing transfer of chromosome-normal embryos [30]. Recently, a clinical study has been conducted on a large multicenter series involving more than 1300 embryos, indicating that embryonic cfDNA originated from both TE cells and ICM cells [31]. Their noninvasive analysis of niPGT-A demonstrates a high concordance with TE biopsy results. Despite promising ongoing research on the clinical concordance results of niPGT-A, there are still concerns regarding the investigation of challenges and improvement of methodology.

3.4 Combination of blastocoel fluid and spent culture medium

Another approach to noninvasive testing is a combination of BF and SCM, which increases the amount of assayable cfDNA. By analyzing blastocysts derived from oocytes not showing evidence of fertilization (no pronuclei visible, 0PN) and oocytes of single pronuclei (1PN), the miPGT-A using BF enriched culture medium shows consistent results of chromosomal aneuploidies compared to TE biopsy PGT-A [32]. In a validation study, Kuznyetsov et al. [33] reported that the accuracy of miPGT-A was not correlated with blastocyst morphological grade. Furthermore, the overall concordance rate of aneuploidy with TE biopsy was approximately 97.8%.

Despite the combination of BF and SCM, the combination of biopsy and noninvasive testing results could also improve the outcomes of assisted pregnancy. Sun et al. [34] have recently reported a similar effectiveness of niPGT-A in assessing embryos with both normal and abnormal karyotypes. Meanwhile, they divided normal embryos into two groups based on the TE biopsy results: euploid in niPGT-A and aneuploid in niPGT-A. By following up on their clinical outcomes, they found that the clinical and ongoing pregnancy rates were higher in the euploid group compared to the aneuploid group, although there was no statistical difference between them. Thus, solely identifying aneuploidy may result in the wastage of embryos due to the high false positives. These findings also suggest that combining biopsy and niPGT-A results could improve the clinical outcome. With the development of noninvasive methodology, it becomes not only an alternative or replacement for present invasive technology but also an encouraging combination to improve ART treatment.

3.5 New perspectives of niPGT-A

During conventional IVF procedure, there is a risk of potential contamination from spermatozoa and cumulus cells attaching to the zona pellucida. As a result, ICSI is typically performed in order for PGT-A to be clinically applicable [35]. However, ICSI is only used to treat couples with severe male factor infertility and is an invasive procedure. Additionally, for patients who did not undergo PGT-A in a fresh cycle and experienced implantation failure or miscarriage due to chromosomal abnormalities, they need to choose existing cryopreserved embryos from IVF for the next transfer cycle. For these patients, selecting embryos based on morphological assessment alone cannot exclude embryos with chromosomal abnormalities. Moreover, PGT-A of frozen embryos requires a series of invasive procedures, including thawing, biopsy, and refreezing. Therefore, if the chromosome ploidy of embryos can be detected by niPGT-A, the need for invasive biopsy will be eliminated, thereby reducing the risk of embryo damage.

A prospective clinical study utilized a quantitative method to detect parental DNA contamination in conventional IVF [36]. The study's findings indicated a low rate of maternal contamination and negligible risk of paternal contamination. PGT-A can be applied to vitrified conventional IVF embryos. Additionally, Xie et al. [37] confirmed that sperm does not impact the niPGT-A results as the current WGA amplification system failed to amplify sperm DNA. They also discovered that niPGT-A from traditional IVF embryos showed high rates of agreement with corresponding TE biopsies, for both fresh and thawed blastocysts. Another research study examined the agreement between niPGT-A results obtained from BF of fresh blastocysts, SCM of thawed embryos, and PGT-A from biopsies [38]. It has shown that SCM has a higher concordance rate and superior diagnostic performance in identifying euploid or aneuploid categories. The performance of SCM is comparable to that of ICSI blastocysts, which gives hope for the potential use of niPGT-A in existing cryopreserved embryos.

Besides the diagnostic potential, combining niPGT-A results with morphological assessment could provide an effective tool to guide frozen-thawed single blastocyst transfer, thus improving the clinical pregnancy outcome [39]. Similarly, for patients experiencing RPL or RIF, selecting euploid embryos through niPGT-A can reduce the miscarriage rate and improve the clinical pregnancy rate [40].

Altogether, these results preliminarily confirm that the niPGT-A approach using SCM from thawing embryos or embryos from conventional IVF has comparable performance to fresh ICSI blastocysts. It could serve as an alternative or complementary approach to embryo biopsy for improving the assessment of preimplantation embryos. Additionally, the potential impact of niPGT-A on infertility treatment may extend to pregnancy outcomes.

3.6 AI models in predicting embryonic ploidy

In addition to noninvasive testing for assessing embryonic chromosomes, static embryo imaging, analyzed by AI algorithms along with clinical parameters, has also shown potential in predicting the ploidy status of preimplantation embryos. Several AI models have been developed, employing different machine learning or deep learning algorithms, such as the random forest classifier (RFC), logistic regression (LR), support vector machine (SVM), k-nearest neighbors (k-NN), and the CNN-based ResNet18 algorithm.

For example, a retrospective study developed an embryo evaluation model, called STORK-A, to predict the ploidy status of embryos [41]. Initially, the study included embryoscope images, blastocyst morphological assessments, morphokinetic parameters, and maternal age and used four different models (XGBoost, k-NN, SVM, and RFC) to determine the importance of variables. Subsequently, it evaluated the accuracy of their deep learning CNN-based algorithm for predicting ploidy. After validating its results in two independent and external datasets, the best-performing models accurately predicted binary aneuploid or euploid embryos with an accuracy of 69.3% and a PPV of 76.1%. For complex aneuploidy, the accuracy was 77.6%. This model is the first to predict complex aneuploidy, expanding the application of AI algorithms in a novel way. Additionally, Jiang et al. [42] demonstrated that combining CNNs of static blastocyst images with clinical characteristics could enhance the accuracy of classifying ploidy status from CNN alone. Therefore, including clinical parameters is essential for improving the predictive performance of AI models.

Besides static images, TL incubators can provide more detailed spatial and temporal information about embryos. By using a deep learning approach to analyze monitoring videos, Paya et al. [43] have identified a noninvasive method for diagnosing aneuploidy. Their model's accuracy ranged from 0.62 to 0.73. Serrano-Novillo et al. [44] discovered that a new parameter during the initial cell cleavage was highly associated with ploidy status. A logistic regression model incorporating various morphokinetic parameters extracted from videos yielded an ROC value of 0.69 for the prediction of chromosome ploidy status. Interestingly, when comparing 12 machine learning models using a multicenter morphokinetic meta-dataset, it was found that mixed effects logistic regression performed better than all other approaches for predicting ploidy [45].

Various AI models have been developed using embryonic morphology from static images or TL videos. In addition to the studies mentioned here, advancements in this field have been discussed in a recently published review paper [46]. The current PGT-A, whether through traditional invasive biopsy or noninvasive screening methods, has several limitations, including financial burdens and delays in result reporting. Due to their ability to quickly make decisions for embryo selection, AI models have the potential to improve clinical pregnancy rates while reducing the cost of each ART cycle.

4. Omics analysis of the SCM to evaluate the developmental potential of embryos

4.1 Molecular markers of embryonic potential in SCM

The evaluation of embryo quality is a crucial factor in determining the success rates of IVF. In order to increase the chances of success, scientists have been constantly searching for reliable methods to predict and select embryos with the highest potential. Currently, the main methods for assessing embrynic potential are morphological assessment and genetic testing for ploidy status. However, these criteria are not able to accurately identify embryos with a high chance of implantation [47]. Therefore, it is of utmost importance to find a noninvasive, reliable, and fast method for assessing embryos. Recently, researchers have started investigating techniques that link molecular markers in SCM with the quality of embryos. This area of research has attracted significant interest due to its potential to enhance the success rates of IVF, decrease the chances of multiple pregnancies, and reduce the expenses associated with infertility treatments.

In recent years, many studies have focused on identifying molecular markers that are closely linked to embryo quality and developmental potential. These markers include metabolites, proteins, and nucleic acids found in SCM. By analyzing these molecules in SCM, researchers can gain insights into the status of the embryo and predict its developmental potential (**Table 1**).

4.1.1 Metabolites

Many researches indicate a strong correlation between the metabolomic characteristics of preimplantation embryos and their potential for development. Changes in the metabolomic profile of SCM can to some extent be used as indicators for selecting embryos and predicting their developmental potential, as well as the clinical outcomes of pregnancy [62]. For example, researchers used NMR spectroscopy to discover that the ratio of acetate to alanine in SCM significantly decreased in embryos with successful implantation, while lactate levels remained similar. By combining these parameters, there is potential to identify a single factor closely linked to implantation potential [48]. In a recent study, Gardner et al. [49] have used microfluorometry to analyze the daily glucose consumption of compacted embryos in SCM. The results showed that transplanted embryos had significantly higher glucose consumption on the 4th and 5th days compared to embryos that did not develop after transplantation. Additionally, it was found that glucose uptake was not dependent on embryo grade.

Studies have found that Raman spectroscopy can be used to detect the metabolic profile of third-day embryo culture medium. This detection can accurately predict the potential of embryos to successfully develop into blastocysts, with an accuracy rate of 73.53% [63]. The use of liquid chromatography-mass spectrometry (LC-MS)

	Markers	Samples	Culture time	Outcome	Reference
METABOLITE	Pyruvat, alanine	48	D3	Implantation outcome	Pudakalakatti et al. [48]
	Glucose	50	D3	Embryo selection	Gardner et al. [49]
Protein	ALCAM, EPHB4, JAM-A/F11R, SELE, CCL24, FAS, PDGF-A, PECAM-1, TIMP4, PON3, CSTB, BMH, vWF	42	D5	Implantation outcome	Freis et al. [50]
	sCD146	126	D2/D3/D5	Embryo selection	Bouvier et al. [51]
	LIF	208	D5	Implantation outcome	Li et al. [52]
	GDF9	399	D3	Implantation outcome	Li et al. [53]
	sHLA-G	539	D3/D5	Implantation outcome	Vani et al. [54]
	HptA1	143	D3	Embryo selection	Montsko et al. [55]
DNA	mtDNA, gDNA	699	D3/D5	Embryo selection	Stigliani et al. [56]
	cfDNA	55	D3	Embryo selection	Pan et al. [57]
RNA	miR-142-3p	36	D3	implantation outcome	Borges et al. [58]
	miR-20a, miR-30c	53	D5/D6	Embryo selection	Capalbo et al. [59]
	miR-191-5p, miR-24-1-5p	50	D5	Implantation outcome	Acuna-Gonzalez et al. [60]
	PROK1	34	D2	Implantation outcome	Alfaidy et al. [61]

Table 1.
Example of molecular markers predicting embryonic development potential.

technology in detecting metabolites in SCM demonstrates the changes of 13 metabolites. This enables noninvasive prediction of the implantation potential of Day 3 embryos [64]. A study has established a method based on metabolomics and data modeling techniques using Fourier transform infrared (FTIR) spectroscopy [65]. This method has successfully identified diverse metabolic profiles in the supernatant of SCM from Day 3 embryos. It has also been able to predict around 60% of non-implanted embryos. Additionally, the study suggests that the oxidative status of SCM may play a role in the developmental health of embryos, specifically in the formation of strong blastocysts and successful implantation. High-quality embryos exhibit a higher level of oxidative metabolism, which results in an increased "oxidative load" on the surrounding culture medium [66].

Consequently, it will be crucial to identify metabolites that can predict the potential for embryonic development, construct reliable predictive models using metabolomics, and integrate them with traditional morphological assessment methods. This will enable genuinely noninvasive, rapid, and accurate evaluations of embryos, making it a pivotal approach for IVF embryo selection in the future.

4.1.2 Proteins

At present, it is possible to accurately and noninvasively determine the proteomic profile of SCM. Targeted studies of specific protein fragments using mass spectrometry and similar methods have also been proven effective in aiding embryo selection [51, 52, 67], thus providing insight into the biochemical condition of the embryos.

Freis et al. [50] utilized the proximity extension assay (PEA) technique for the first time to analyze individual blastocyst culture media influenced by their potential for implantation. They examined the expression of various proteins that could serve as potential biomarkers for noninvasive assessment of individual blastocysts. The expression levels of ALCAM, EPHB4, JAM-A/F11R, SELE, CCL24, FAS, PDGF-A, PECAM-1, TIMP4, PON3, CSTB, BMH, and vWF were found to be significantly higher in the SCM of successfully implanted blastocysts compared to nonimplanted blastocysts. Li et al. [52] discovered that there is a significant increase in LIF levels in the SCM of embryos in the pregnant group compared to the nonpregnant group. The concentration of GDF9 in SCM is correlated with embryo quality and clinical outcomes [53]. Specifically, in 94% of couples, nontransferable embryos show a higher GDF9 concentration in the culture medium compared to transferable embryos. Interestingly, the levels of sHLA-G are positively correlated with blastocyst grading. In the group with live births, the levels of sHLA-G are significantly higher than those in the group without live births, suggesting the potential to predict successful pregnancy outcomes [54]. Montsko et al. [55] employed liquid chromatography combined with mass spectrometry to analyze HptA1. Their research showed that the degree of Hpt cleavage and the concentration of HptA1 in the culture medium can accurately predict the results of embryo transplantation. Montsko et al. [67] first demonstrated that sCD146 can be detected in SCM. Additionally, a high concentration of sCD146 is linked to a decreased likelihood of embryo implantation.

4.1.3 Nucleic acids

Cell-free nucleic acids in SCM consist of cfDNA and cell-free RNA (cfRNA). Studies have revealed that in the culture medium of embryos cultured for 3 days, the ratio of mitochondrial DNA (mtDNA) to genomic DNA (gDNA) is correlated with the outcome of embryo transplantation [56]. Besides, embryos that successfully develop into blastocysts exhibit a significantly higher mtDNA/gDNA ratio in the culture medium compared to those that do not develop into blastocysts. Pan et al. [57] assessed the correlation between cfDNA fragment patterns in follicular fluid and SCM of patients with subsequent embryo grading, demonstrating for the first time a significant association between cfDNA in follicular fluid and SCM with both embryo occurrence and embryo grading.

Research has found that microRNAs (miRNAs) are expressed during embryo cell division and genome activation processes. Additionally, small RNAs can be secreted into the SCM [68]. Several researchers have also identified miRNA molecular markers associated with embryo implantation outcomes and pregnancy outcomes [58–60, 69].

In the subgroup of negative implantation in the SCM, there was a notable rise in the expression of miR-142-3p. This observation confirms the differential expression of miR-142-3p in human embryos as they progress to the blastocyst stage, depending on their implantation status [58]. A study conducted a comprehensive characterization of the miRNAs of SCM from euploid blastocysts that were either implanted or unimplanted. The study highlighted two specific miRNAs, miR-20a and miR-30c, which exhibited higher concentrations in the implanted blastocysts. Additionally, *in silico* predictions suggested that these miRNAs may play a role in embryonic implantation [59]. Kamijo et al. [69] identified 53 differentially expressed miRNAs in the culture medium of the pregnancy and nonpregnancy groups. By utilizing logistic regression analysis, a high-quality blastocyst prediction model was generated with 8 miRNAs, resulting in an average accuracy of 0.82. Another research indicates that miR-191-5p in SCM could potentially serve as a positive biomarker for pregnancy, while miR-24-1-5p may indicate a poor prognosis [60]. This former miRNA regulates IGF2BP-1 and IGF2R, which are associated with the implantation window. Conversely, miR-24-1-5p may be associated with a negative prognosis for human embryo development.

At the cellular level, the concentration of the *PROK1* gene in follicular fluid and SCM was determined using real-time quantitative amplification technology. This study found a correlation between the *PROK1* gene content and the success of embryo implantation [61]. It is speculated that high levels of *PROK1* may support a strategy of transferring a single embryo, while low levels of *PROK1* may serve as an additional criterion for extending embryo culture or performing double embryo transfer in order to enhance IVF outcomes.

4.2 AI models in predicting developmental potential

In addition to the noninvasive testing of molecular markers in SCM, AI models have also been applied to predict clinical pregnancy by utilizing clinical treatment information, along with embryonic images or TL videos.

Researchers used a semantic segmentation neural network model to measure embryonic morphometric parameters, including blastocyst size, ICM size, and ICM shape. The study supported the correlation between larger blastocyst size and increased potential for implantation [70]. Using colored blastocyst images, a novel AI system called FiTTE has the potential to offer more accurate predictions for clinical pregnancy compared to the conventional Gardner scoring system [71]. A research team has developed iDAScore v2.0, a fully automated deep learning model, using an analysis of TL sequences from a diverse and extensive dataset collected across 22 IVF centers in order to rank embryos. The accuracy rate for predicting implantation outcome varies between 0.621 and 0.707, depending on the day of transfer [72]. In a subsequent retrospective observational study carried out at two additional IVF centers, iDAScore v2.0 was able to distinguish between embryos that led to live birth or no live birth. The AUC for embryos transferred at Day 2 was 0.627, while for embryos at Day 3, the value was 0.607. These findings indicate that iDAScore v2.0 holds some predictive value for live birth [73]. Based on six nonredundant morphodynamical features of TL images, Amitai et al. [74] developed a decision-support tool for identifying embryos at high risk of first-trimester miscarriage. Prioritizing embryos based on their predicted risk of miscarriage, in combination with their implantation potential, is expected to improve live birth rates and shorten the time to pregnancy. Recently, Liu et al. [75] have trained an AI model using CNN to process blastocyst images and a multilayer perceptron to process clinical features. Their work suggested that including clinical features of patient couples along with blastocyst images increased the accuracy of predicting live birth. They achieved an AUC of 0.77. Meanwhile, Duval et al. [76] trained algorithms on multicentric clinical data using a hybrid model. The average AUC of predicting fetal heart rate during pregnancy significantly increased when using their hybrid model compared to the algorithms solely based on the TL videos. This AI model can assist embryologists in various clinical scenarios.

In summary, the AI models can predict the potential of embryonic development in various aspects. On average, the AI models achieved a median accuracy of 77.8% (range 68–90%) in predicting clinical pregnancy. When combining both TL images or videos and clinical information inputs, the AI models demonstrated a higher median accuracy of 81.5% (range 67–98%) [10]. The evolving AI models hold significant promise for the IVF field and embryo selection.

5. Discussion

Noninvasive testing brings new possibilities in evaluating preimplantation embryos, such as examining their morphology, conducting genetic testing for aneuploidy, and predicting their developmental potential. While the substances released into the BF and SCM are not confined to cfDNA, other omics analyses of the SCM can also enable the simultaneous assessment of chromosome ploidy and provide additional information. A recent study has successfully utilized the DNA methylation approach of cfDNA in the SCM to accurately detect chromosome aneuploidy of embryos, trace the cellular origins of the SCM, and confirm the suspicion of cumulus contamination in the SCM. This breakthrough aims to establish a DNA methylation-based niPGT-A method [77]. The development of a methodology to achieve simultaneous quantification and characterization of cfRNA [78] also provides opportunities to comprehensively profile cfRNA in SCM and discover additional molecular markers associated with developmental potential.

In addition, instead of treating the noninvasive parameters for embryo selection mentioned above as separate variables related to the normality and quality of embryos, it would be wise to consider them collectively. Utilizing AI models along with a comprehensive set of information for each embryo can result in a more precise prediction. For instance, integrating machine learning with metabolomic and embryologic data could enhance the accuracy of predicting the potential for embryo implantation [79]. Developing a grading system with AI models using omics data has the potential to enhance the quality and efficiency of IVF procedures, resulting in real-time clinical benefits for patients.

6. Conclusion

Although the invasive embryo biopsy technology for preimplantation genetic testing (PGT) has been widely used for embryo selection in human IVF practice, it is detrimental to embryos. Thus, searching a method of the noninvasive testing of preimplantation embryos has gradually become a hot project in ART. Here, we reviewed the current progression of noninvasive embryo testing technologies, including morphokinetics from TL images, miPGT-A of BF, niPGT-A of SCM, combination analysis of BF and SCM, and molecules in SCM and clinical information associated with developmental potential of embryos. The combination of these methods will be helpful to achieve comprehensive assessment of preimplantation embryos so that the normal embryos could be selected for transfer. It is important that we think that integrating AI technology into these noninvasive embryo assessments based on vast amounts of different data types for each embryo will help us to more accurately select normal genetic embryos and with great developmental potential. This combined technology will make ART more effective and cheaper for patients undergoing infertility treatment.

Acknowledgements

This project is supported by the Guangdong Basic and Applied Basic Research Foundation (No. 2023A1515110478).

Conflict of interest

The authors declare no conflict of interest.

Author details

Qing Zhou* and Yutong Wang
BGI Research, Shenzhen, China

*Address all correspondence to: zhouqing1@genomics.cn

References

[1] Njagi P et al. Financial costs of assisted reproductive technology for patients in low- and middle-income countries: A systematic review. Human Reproduction Open. 2023;**2023**(2):hoad007

[2] Zegers-Hochschild F et al. The international glossary on infertility and fertility care, 2017. Human Reproduction. 2017;**32**(9):1786-1801

[3] Bashiri A, Halper KI, Orvieto R. Recurrent implantation failure-update overview on etiology, diagnosis, treatment and future directions. Reproductive Biology and Endocrinology. 2018;**16**(1):121

[4] Simon A, Laufer N. Assessment and treatment of repeated implantation failure (RIF). Journal of Assisted Reproduction and Genetics. 2012;**29**(11):1227-1239

[5] Kirkegaard K et al. Choosing the best embryo by time lapse versus standard morphology. Fertility and Sterility. 2015;**103**(2):323-332

[6] Gardner DK, Sakkas D. Assessment of embryo viability: The ability to select a single embryo for transfer—A review. Placenta. 2003;**24**:S5-S12

[7] Baxter Bendus AE et al. Interobserver and intraobserver variation in day 3 embryo grading. Fertility and Sterility. 2006;**86**(6):1608-1615

[8] Dirvanauskas D et al. Embryo development stage prediction algorithm for automated time lapse incubators. Computer Methods and Programs in Biomedicine. 2019;**177**:161-174

[9] Glatstein I, Chavez-Badiola A, Curchoe CL. New frontiers in embryo selection. Journal of Assisted Reproduction and Genetics. 2023;**40**(2):223-234

[10] Salih M et al. Embryo selection through artificial intelligence versus embryologists: A systematic review. Human Reproduction Open. 2023;**2023**(3):hoad031

[11] Khosravi P et al. Deep learning enables robust assessment and selection of human blastocysts after in vitro fertilization. npj Digital Medicine. 2019;**2**(1):21

[12] Loewke K et al. Characterization of an artificial intelligence model for ranking static images of blastocyst stage embryos. Fertility and Sterility. 2022;**117**(3):528-535

[13] Minasi MG et al. Correlation between aneuploidy, standard morphology evaluation and morphokinetic development in 1730 biopsied blastocysts: A consecutive case series study. Human Reproduction. 2016;**31**(10):2245-2254

[14] Sahoo T et al. Comprehensive genetic analysis of pregnancy loss by chromosomal microarrays: Outcomes, benefits, and challenges. Genetics in Medicine. 2017;**19**(1):83-89

[15] Neal SA et al. Preimplantation genetic testing for aneuploidy is cost-effective, shortens treatment time, and reduces the risk of failed embryo transfer and clinical miscarriage. Fertility and Sterility. 2018;**110**(5):896-904

[16] Rubio C et al. In vitro fertilization with preimplantation genetic diagnosis for aneuploidies in advanced maternal age: A randomized,

controlled study. Fertility and Sterility. 2017;**107**(5):1122-1129

[17] Cinnioglu C et al. A systematic review of noninvasive preimplantation genetic testing for aneuploidy. Fertility and Sterility. 2023;**120**(2):235-239

[18] Palini S et al. Genomic DNA in human blastocoele fluid. Reproductive Biomedicine Online. 2013;**26**(6):603-610

[19] Huang L et al. Noninvasive preimplantation genetic testing for aneuploidy in spent medium may be more reliable than trophectoderm biopsy. Proceedings of the National Academy of Sciences. 2019;**116**(28):14105

[20] Gianaroli L et al. Blastocentesis: A source of DNA for preimplantation genetic testing. Results from a pilot study. Fertility and Sterility. 2014;**102**(6): 1692-9 e6

[21] Tobler KJ et al. Blastocoel fluid from differentiated blastocysts harbors embryonic genomic material capable of a whole-genome deoxyribonucleic acid amplification and comprehensive chromosome microarray analysis. Fertility and Sterility. 2015;**104**(2):418-425

[22] Stigliani S et al. Mitochondrial DNA content in embryo culture medium is significantly associated with human embryo fragmentation. Human Reproduction. 2013;**28**(10):2652-2660

[23] Farra C, Choucair F, Awwad J. Non-invasive pre-implantation genetic testing of human embryos: An emerging concept. Human Reproduction. 2018;**33**(12):2162-2167

[24] Hammond ER, Shelling AN, Cree LM. Nuclear and mitochondrial DNA in blastocoele fluid and embryo culture medium: Evidence and potential clinical use. Human Reproduction. 2016;**31**(8):1653-1661

[25] Xu J et al. Noninvasive chromosome screening of human embryos by genome sequencing of embryo culture medium for in vitro fertilization. Proceedings of the National Academy of Sciences of the United States of America. 2016;**113**(42):11907-11912

[26] Capalbo A et al. Diagnostic efficacy of blastocoel fluid and spent media as sources of DNA for preimplantation genetic testing in standard clinical conditions. Fertility and Sterility. 2018;**110**(5):870-879 e5

[27] Hanson BM et al. Noninvasive preimplantation genetic testing for aneuploidy exhibits high rates of deoxyribonucleic acid amplification failure and poor correlation with results obtained using trophectoderm biopsy. Fertility and Sterility. 2021;**115**(6):1461-1470

[28] Yin B et al. Validation of preimplantation genetic tests for aneuploidy (PGT-A) with DNA from spent culture media (SCM): Concordance assessment and implication. Reproductive Biology and Endocrinology. 2021;**19**(1):41

[29] Jiao J et al. Minimally invasive preimplantation genetic testing using blastocyst culture medium. Human Reproduction. 2019;**34**(7):1369-1379

[30] Yang P et al. Downregulated miR-451a as a feature of the plasma cfRNA landscape reveals regulatory networks of IL-6/IL-6R-associated cytokine storms in COVID-19 patients. Cellular & Molecular Immunology. 2021;**18**(4):1064-1066

[31] Rubio C et al. Multicenter prospective study of concordance between embryonic cell-free DNA

and trophectoderm biopsies from 1301 human blastocysts. American Journal of Obstetrics and Gynecology. 2020;**223**(5):751 e1-751 e13

[32] Zhao H et al. The aneuploidy testing of blastocysts developing from 0PN and 1PN zygotes in conventional IVF through TE-biopsy PGT-A and minimally invasive PGT-A. Frontiers in Reproductive Health. 2022;**4**:966909

[33] Kuznyetsov V et al. Minimally invasive cell-free human embryo aneuploidy testing (miPGT-A) utilizing combined spent embryo culture medium and blastocoel fluid -towards development of a clinical assay. Scientific Reports. 2020;**10**(1):7244

[34] Sun B-L et al. Effectiveness of non-invasive chromosomal screening for normal karyotype and chromosomal rearrangements. Frontiers in Genetics. 2023;**14**:1036467

[35] De Rycke M et al. ESHRE PGT consortium and SIG embryology good practice recommendations for polar body and embryo biopsy for PGT†. Human Reproduction Open. 2020;**2020**(3):hoaa020

[36] Dong Y et al. Preimplantation genetic testing for human blastocysts with potential parental contamination using a quantitative parental contamination test (qPCT): An evidence-based study. Reproductive Biomedicine Online. 2023;**46**(1):69-79

[37] Xie P et al. Non-invasive preimplantation genetic testing for conventional IVF blastocysts. Journal of Translational Medicine. 2022;**20**(1):396

[38] Shi W et al. Ploidy testing of blastocoel fluid for screening may be technically challenging and more invasive than that of spent cell culture media. Frontiers in Physiology. 2022;**13**:794210

[39] Chen R et al. Clinical application of noninvasive chromosomal screening for elective single-blastocyst transfer in frozen-thawed cycles. Journal of Translational Medicine. 2022;**20**(1):553

[40] Xi H et al. Noninvasive chromosome screening for evaluating the clinical outcomes of patients with recurrent pregnancy loss or repeated implantation failure. Frontiers in Endocrinology. 2022;**13**:896357

[41] Barnes J et al. A non-invasive artificial intelligence approach for the prediction of human blastocyst ploidy: A retrospective model development and validation study. Lancet Digit Health. 2023;**5**(1):e28-e40

[42] Jiang VS et al. The use of voting ensembles to improve the accuracy of deep neural networks as a non-invasive method to predict embryo ploidy status. Journal of Assisted Reproduction and Genetics. 2023;**40**(2):301-308

[43] Paya E et al. Deep learning system for classification of ploidy status using time-lapse videos. F&S Science. 2023;**4**(3):211-218

[44] Serrano-Novillo C, Uroz L, Márquez C. Novel time-lapse parameters correlate with embryo ploidy and suggest an improvement in non-invasive embryo selection. Journal of Clinical Medicine. 2023;**12**(8):2983

[45] Bamford T et al. A comparison of 12 machine learning models developed to predict ploidy, using a morphokinetic meta-dataset of 8147 embryos. Human Reproduction. 2023;**38**(4):569-581

[46] Jiang VS, Bormann CL. Noninvasive genetic screening: Current advances in artificial intelligence for embryo ploidy prediction. Fertility and Sterility. 2023;**120**(2):228-234

[47] Jiang XM et al. Does conventional morphological evaluation still play a role in predicting blastocyst formation? Reproductive Biology and Endocrinology. 2022;**20**(1):68

[48] Pudakalakatti SM et al. NMR studies of preimplantation embryo metabolism in human assisted reproductive techniques: A new biomarker for assessment of embryo implantation potential. NMR in Biomedicine. 2013;**26**(1):20-27

[49] Gardner DK et al. Glucose consumption of single post-compaction human embryos is predictive of embryo sex and live birth outcome. Human Reproduction. 2011;**26**(8):1981-1986

[50] Freis A et al. Non-invasive embryo assessment: Altered individual protein profile in spent culture media from embryos transferred at day 5. Reproductive Sciences. 2021;**28**(7):1866-1873

[51] Bouvier S et al. Soluble CD146, an innovative and non-invasive biomarker of embryo selection for in vitro fertilization. PLoS One. 2017;**12**(3):e0173724

[52] Li Z et al. LIF in embryo culture medium is a predictive marker for clinical pregnancy following IVF-ET of patients with fallopian tube obstruction. Journal of Reproductive Immunology. 2020;**141**:103164

[53] Li J et al. GDF9 concentration in embryo culture medium is linked to human embryo quality and viability. Journal of Assisted Reproduction and Genetics. 2022;**39**(1):117-125

[54] Vani V et al. Soluble human leukocyte antigen-G is a potential embryo viability biomarker and a positive predictor of live-births in humans. American Journal of Reproductive Immunology. 2021;**86**(6):e13499

[55] Montskó G et al. Alpha-1 chain of human haptoglobin as viability marker of in vitro fertilized human embryos: Information beyond morphology. Systems Biology in Reproductive Medicine. 2019;**65**(2):174-180

[56] Stigliani S et al. Mitochondrial DNA in day 3 embryo culture medium is a novel, non-invasive biomarker of blastocyst potential and implantation outcome. Molecular Human Reproduction. 2014;**20**(12):1238-1246

[57] Pan M et al. The integrity of cfDNA in follicular fluid and spent medium from embryo culture is associated with embryo grade in patients undergoing in vitro fertilization. Journal of Assisted Reproduction and Genetics. 2021;**38**(12):3113-3124

[58] Borges E Jr et al. miR-142-3p as a biomarker of blastocyst implantation failure - A pilot study. JBRA Assisted Reproduction. 2016;**20**(4):200-205

[59] Capalbo A et al. MicroRNAs in spent blastocyst culture medium are derived from trophectoderm cells and can be explored for human embryo reproductive competence assessment. Fertility and Sterility. 2016;**105**(1):225-235 e1-3

[60] Acuña-González RJ et al. MiR-191-5p is upregulated in culture media of implanted human embryo on day fifth of development. Reproductive Biology and Endocrinology. 2021;**19**(1):109

[61] Alfaidy N et al. PROK1 level in the follicular microenvironment: A new noninvasive predictive biomarker of embryo implantation. The Journal of Clinical Endocrinology and Metabolism. 2016;**101**(2):435-444

[62] Drabkova P et al. Changes in the levels of selected metabolites in the culture medium as a possible tool for the embryo selection in assisted reproduction. Ceská Gynekologie. 2015;**80**(2):135-139

[63] Zheng W et al. Non-invasive metabolomic profiling of embryo culture medium using Raman spectroscopy with deep learning model predicts the blastocyst development potential of embryos. Frontiers in Physiology. 2021;**12**:777259

[64] Liang R et al. Prediction model for day 3 embryo implantation potential based on metabolites in spent embryo culture medium. BMC Pregnancy and Childbirth. 2023;**23**(1):425

[65] Figoli CB et al. A robust metabolomics approach for the evaluation of human embryos from in vitro fertilization. Analyst. 2021;**146**(20):6156-6169

[66] Alegre L et al. Novel noninvasive embryo selection algorithm combining time-lapse morphokinetics and oxidative status of the spent embryo culture medium. Fertility and Sterility. 2019;**111**(5):918-927 e3

[67] Montskó G et al. Noninvasive embryo viability assessment by quantitation of human haptoglobin alpha-1 fragment in the in vitro fertilization culture medium: An additional tool to increase success rate. Fertility and Sterility. 2015;**103**(3):687-693

[68] Kropp J, Salih SM, Khatib H. Expression of microRNAs in bovine and human pre-implantation embryo culture media. Frontiers in Genetics. 2014;**5**:91

[69] Kamijo S et al. MicroRNAs secreted by human preimplantation embryos and IVF outcome. Reproductive Biology and Endocrinology. 2022;**20**(1):130

[70] Fruchter-Goldmeier Y et al. An artificial intelligence algorithm for automated blastocyst morphometric parameters demonstrates a positive association with implantation potential. Scientific Reports. 2023;**13**(1):14617

[71] Enatsu N et al. A novel system based on artificial intelligence for predicting blastocyst viability and visualizing the explanation. Reproductive Medicine and Biology. 2022;**21**(1):e12443

[72] Theilgaard Lassen J et al. Development and validation of deep learning based embryo selection across multiple days of transfer. Scientific Reports. 2023;**13**(1):4235

[73] Ahlström A et al. Correlations between a deep learning-based algorithm for embryo evaluation with cleavage-stage cell numbers and fragmentation. Reproductive Biomedicine Online. 2023;**47**(6):103408

[74] Amitai T et al. Embryo classification beyond pregnancy: Early prediction of first trimester miscarriage using machine learning. Journal of Assisted Reproduction and Genetics. 2023;**40**(2):309-322

[75] Liu H et al. Development and evaluation of a live birth prediction model for evaluating human blastocysts from a retrospective study. eLife. 2023;**12**:e83662

[76] Duval A et al. A hybrid artificial intelligence model leverages multi-centric clinical data to improve fetal heart rate pregnancy prediction across time-lapse systems. Human Reproduction. 2023;**38**(4):596-608

[77] Chen Y et al. DNA methylome reveals cellular origin of cell-free DNA in spent medium of human preimplantation embryos. The Journal of Clinical Investigation. 2021;**131**(12):e146051

[78] Liu Z et al. Polyadenylation ligation-mediated sequencing (PALM-Seq) characterizes cell-free coding and non-coding RNAs in human biofluids. Clinical and Translational Medicine. 2022;**12**(7):e987

[79] Cheredath A et al. Combining machine learning with metabolomic and embryologic data improves embryo implantation prediction. Reproductive Sciences. 2023;**30**(3):984-994

Chapter 4

The Research Progress of Correcting Pathogenic Mutations by Base Editing

Guanglei Li, Yu Zhang and Xingxu Huang

Abstract

Over 6500 Mendelian disorders have been documented, with approximately 4500 genes linked to these conditions. The majority of inherited diseases present in childhood and, currently, lack effective treatments, which imposes significant economic and psychological burdens on families and society. Gene editing, particularly base editing, offers an effective and safe strategy for repairing pathogenic point mutations. It has the potential to become a treatment, even a cure, for rare diseases. Currently, multiple gene editing-related drugs have entered clinical trials. In this chapter, we summarize the various gene editing systems, including CRISPR/Cas, base editing, and prime editing. We then focus on the current research progress of base editing in correcting pathogenic mutations. This includes applications such as building animal models, correcting mutations in various diseases, germline cell editing, delivery methods, and approved clinical trials. Finally, we discuss current challenges related to delivery methods, efficiency, precision, and cost.

Keywords: Mendelian disorders, economic and psychological burdens, pathogenic point mutations, base editing, clinical trials

1. Introduction

On November 16 and December 8 in 2023, the Medicines and Healthcare products Regulatory Agency (MHRA) and the Food and Drug Administration (FDA), respectively, authorized the world's first gene editing product, Casgevy, for the treatment of sickle cell disease and transfusion-dependent β-thalassemia [1]. It has been 11 years since the initial report on the CRISPR/Cas system's ability to cut double-stranded DNA [2]. Casgevy is classified as the first generation of genome editing, utilizing CRISPR/Cas9 to create a double-stranded break (DSB) at the erythroid enhancer region of the BCL11A gene [3]. The authorization of Casgevy will set a precedence for the treatment of inherited diseases through gene editing methods. Although Casgevy is the first approved gene editing product, it may not be the most advanced. The CRISPR/Cas system's ability to induce DSBs, base editing, prime editing, and epigenome editing allows for more precise and versatile DNA editing, paving the way for CRISPR 2.0 [4].

In the post-genome era, it has become increasingly important to clarify the relationship between genotype and phenotype. Additionally, we have confirmed that many inherited diseases are caused by gene mutations. As of December 15, 2023, 7461 types of diseases have been reported, and they are associated with 4871 genes (https://www.omim.org/statistics/geneMap). 6404 of these diseases are single-gene Mendelian disorders and traits. With the widespread availability of sequencing technology, this number will continue to grow. Rare genetic diseases affect millions of people worldwide, and many of them lack effective treatment options. Some genetic diseases can significantly reduce quality of life and lead to premature death [5], especially for developing countries. Gene therapy, including gene addition and gene editing, offers a potential way to correct defective genes [6]. Gene addition using viruses as delivery methods has been proposed for decades, and recently, some products have been approved. However, gene addition is mainly suitable for loss-of-function diseases, while gene editing is more suitable for gain-of-function diseases. The combination of gene addition and gene editing with cell therapy provides an alternative treatment for inherited diseases.

Zinc finger nucleases (ZFN), transcription activator-like effector nucleases (TALEN), and clustered regularly interspaced short palindromic repeats (CRISPR)-CRISPR-associated protein (Cas) systems, particularly CRISPR/Cas9, have significantly advanced the field of gene therapy. All of these methods achieve site-specific gene editing, and several related products using these technologies are currently in clinical trialss [7, 8]. Gene-editing technologies are used not only to correct pathogenic mutations for clinical applications but also to modify immune cells, such as chimeric antigen receptor T (CAR-T) cells [9]. In this chapter, we primarily discuss the application of base editing. In the first section, we present an overview of the development of CRISPR/Cas systems, base editing, and prime editing. We also discuss the potential disadvantage of double-stranded breaks. In the second section, we review the progress made in using base editing to correct inherited pathogenic mutations and present an overview of some preclinical studies. In the third section, we highlight recent clinical trials that have employed base editing.

2. Target gene editing mediated by CRISPR/Cas system

CRISPR/Cas systems are commonly found in bacteria and archaea, where they defend against foreign nucleic acids. CRISPR/Cas systems are typically classified into two classes, with six subtypes: I, III, and IV in class I and II, V, and VI in class II. Recently, Zhang's group has identified the VII subtype using a deep terascale clustering method [10]. Among the subtypes, CRISPR/Cas9 and CRISPR/Cas12 from type II are commonly utilized. The CRISPR/Cas system performs the function of targeting, and the editing results depend on the repair pathway following double-stranded break or on the functional protein fused to Cas. The CRISPR/Cas system has recently found widespread applications in life science, agriculture, and medicine. Numerous other systems have been developed based on CRISPR/Cas, including CRISPRa, CRISPRi, epigenome editing, base editing (BE), and prime editing (PE). In 2020, Emmanuelle Charpentier and Jennifer A. Doudna were awarded the Nobel Prize in Chemistry for the development of a method for genome editing.

2.1 Gene editing and regulation mediated by CRISPR/Cas

The CRISPR/Cas9 system essentially consists of two components, Cas9 and single guide RNA (sgRNA). The two combine to form a complex that then seeks out and binds to the target site. Initially, the most prevalent use of CRISPR/Cas9 was in disrupting target genes, resulting in insertions or deletions (indels) mediated by nonhomologous end joining (NHEJ). Subsequently, the precise genotypes could be derived with the aid of homology-directed repair. Moving beyond basic DNA editing, developments were made in creating systems like CRISPRa, CRISPRi, and epigenome editing. These do not modify the DNA sequence itself. Fundamentally, all of them are fusion proteins that combine different functionally active proteins with a deactivated Cas protein [11, 12].

2.2 Base editing provides a robust tool for making point mutation

The safety of gene editing products is a critical concern for clinical applications. Double-strand breaks are among the most lethal forms of DNA damage. Prior research has indicated that DSB-based gene editing may lead to the deactivation of p53 expression in responsive cells [13]. Moreover, DSBs can occasionally cause the loss of the targeted chromosome in human primary T cells [14]. The risk escalates when multiple sites are edited within a single cell, potentially leading to genomic rearrangements or fragment deletion [15]. Therefore, the risks associated with DSBs need to be carefully considered, especially in the context of gene therapy products [16].

Base editing, first introduced by Liu's group [17], is a technique that combines a deaminase enzyme with a catalytically impaired Cas nuclease. This method allows for efficient and precise base conversions without inducing double-strand breaks (DSBs) or necessitating a donor DNA template. In 2017, base editing technology was heralded as the "Breakthrough of the Year". Among the most utilized base editing systems are cytosine base editors (CBEs), which facilitate C-to-T conversions, and adenine base editors (ABEs), which enable A-to-G conversions. Other base editors, such as C•G-to-G•C base editors (CGBEs) that mediate C-to-G conversions, have also been developed [18]. Base editors incorporate the CRISPR/Cas system and a deaminase enzyme. Various combinations of Cas proteins and deaminases result in distinct editors, each carrying different editing windows or protospacer adjacent motif (PAM) requirements. Continual optimization of base editing systems allows for effective point mutation induction. For instance, Diorio et al. reported using a CBE to introduce point mutations in primary T cells with over 90% efficiency [9]. This manipulation effectively silenced gene expression without inducing DSBs.

Because deaminase binds randomly to ssDNA or ssRNA, base editing systems have been reported to induce sgRNA-independent off-target effects. For CBE, off-target effects occur in both DNA and RNA, while for ABE, off-target effects are mainly found in RNA. Numerous methods have been developed to eliminate these off-target effects, including making mutations in the deaminase, embedding deaminase [19], and using high-fidelity Cas9 and the transformer base editor (tBE) [20]. In addition to engineering base editors, the delivery method can also reduce off-target effects, such as electroporation with CBE protein-replacing plasmids.

Base editing has the potential to introduce point mutations, but its application scope is extensive. One of the most significant applications is correcting pathogenic mutations, as the majority of these are point mutations, making them suitable candidates for base editing. Furthermore, base editing can be used to knock out target genes by introducing stop codons into the coding sequence (CDS), disrupting splicing sites, and destroying start codons [21]. Currently, there are four clinical trials underway using base editing systems, and numerous preclinical studies are in progress [22].

In addition to DNA base editing, RNA base editing and mitochondrial base editing are also emerging technologies that are worth further exploration [23, 24]. By fusing wild-type adenosine deaminase acting on RNA (ADAR) with Cas13, A-to-I base editing in RNA can be achieved, and the evolved ADAR has the ability to convert C-to-U. Even without Cas, native ADAR can be recruited by engineered RNAs to perform A-to-I editing [24]. DddA-derived cytosine base editors (DdCBEs) can convert C-to-T in mitochondrial DNA, while split TALE deaminases (TALEDs) can convert A-to-G in mitochondrial DNA [23].

2.3 Prime editing broadens the editing scope of base editing

Prime editing, first reported by Liu's group in 2019, involves the fusion of an engineered reverse transcriptase to a catalytically impaired Cas9 endonuclease [25]. Additionally, the guide RNA for prime editing (PE) differs from other editing systems. The PE system utilizes a prime editing guide RNA (pegRNA) that not only specifies the target site but also encodes the desired edit. Prime editing is capable of inducing all 12 possible base-to-base conversions, as well as the insertion or deletion of short sequences. Since its initial publication, various laboratories worldwide have contributed to furthering this groundbreaking technology. With continuous optimization, the PE system has evolved from PE1 to PE6 [26]. Prime editing has also been successfully employed in correcting pathogenic mutations, with numerous preclinical studies reporting its effectiveness [27].

3. Animal disease models mediated by base editing

Animal models are crucial for disease research, and CRISPR/Cas9 has become a widely used tool in creating animal disease models [28]. Three main methods exist for producing animal models: embryo microinjection, somatic cell nuclear transfer, and *in situ* injection. These methods have been commonly employed with CRISPR/Cas9. Base editing, like CRISPR/Cas9, offers a similar approach, and animal models based on base editing have been reported.

Kim et al. delivered the cytidine base editing system into mouse embryos to target the *Dmd* gene, resulting in *Dmd* mutant mice that rarely expressed dystrophin protein. For the *Tyr* gene, newborn pups exhibited an albino phenotype in their eyes [29]. Using a similar method, the same group generated *Tyr* mutant mice with an albino phenotype using adenine base editing in 2018 [30]. Beyond rodent animals, base editing has also been applied in zebrafish and nonhuman primates. With the CBE4max-SpRY CBE variant, which recognizes nearly all PAM sequences, Rosello et al. created a zebrafish model for melanoma predisposition [31]. In 2023, Zheng et al. simulated a pathogenic site of the *tsr2* ribosome maturation factor (tsr2) in zebrafish by injecting a mixture of zSpRY-ABE8e mRNA and sgRNA into embryos. The mutant zebrafish exhibited a phenotype of smaller eyes and a swollen pericardium [32]. Wang et al. generated monkey models with Hutchinson-Gilford progeria syndrome (HGPS). They microinjected RNA of CBE into the embryos, and five out of six newborn monkeys carried the mutation. These HGPS monkeys displayed typical HGPS phenotypes including growth retardation, bone alterations, and vascular abnormalities [33].

The RAG1 and IL2RG genes play crucial roles in the immune system, and mutations in these genes can lead to severe combined immunodeficiency (SCID). Zheng et al. developed an immunodeficient monkey model using the CBE4max system. CBE induced a stop codon at the coding sequence of these two genes, resulting in immunodeficient monkeys that showed the ability to foster tumor growth [34].

Base editing holds significant potential in creating animal models. The primary approach involves microinjecting the mRNA- or RNP-encoding base editing into embryos. In contrast to NHEJ-induced knockout methods, base editing allows for the straightforward generation of desired animals without mosaicism at the F0 generation. However, base editing has limitations in terms of its inability to introduce insertions and deletions. As such, HDR, BE, and PE are complementary techniques for creating various animal disease models.

4. Progress of base editing in correcting inherited pathogenic mutations

Rare genetic diseases affect millions of people worldwide and are often caused by mutations in the genome. Gene therapy offers a fundamental approach to curing these rare diseases. A survey of 1430 researchers agreed that gene therapies will greatly benefit the treatment of patients with rare genetic diseases [5]. While gene addition is useful for loss-of-function mutations, and some approved gene therapy products rely on gene addition, the main reason for its effectiveness is the lack of appropriate techniques to fix defective genes. Correcting these defective genes is a more logical approach for rare genetic diseases. The CRISPR/Cas system provides an effective method to change gene mutations, and the base editing system is the best choice for point mutations. Taking into account efficiency and safety, base editing may be more successful in treating rare diseases. Here, we summarize the research progress of correcting pathogenic mutations using base editing (**Table 1**).

4.1 *In vivo* and *ex vivo* gene therapy

Delivery methods play a crucial role in achieving efficient gene therapy. Gene therapy can be categorized into two types: *in vivo* and *ex* vivo [70, 71]. Recently, the majority of studies have focused on *ex vivo* gene therapy. In this approach, cells such as hematopoietic stem and progenitor cells (HSPCs) are extracted from patients and then subjected to the delivery of the editing system. For genetic engineering using base editing, nucleofection with mRNA combined with sgRNA is the primary method [22]. The challenge lies in producing large-scale proteins of base editors, which hinders the application of RNP for base editing purposes. On the other hand, for *in vivo* gene therapy, adeno-associated virus (AAV) and lipid nanoparticles (LNP) are considered as the main delivery methods for delivering the base editing system to target cells [71] (**Figure 1**). Enhancing efficiency and safety in delivery methods will significantly advance progress in *in vivo* gene therapy.

4.2 Hematological disorder

The hematologic system plays a crucial role in multiple vital functions within our bodies, including delivering oxygen and defending against offensive pathogens.

Disease	Gene	Base editor	*In vitro*/*in vivo*	Delivery method	Reference
Fanconi anemia	FANCA	CBE, ABE	FA patient fibroblasts and LCL	Electroporation	Sipe [35]
Fanconi Anemia	FANCA	ABE	HSPC	nucleofection	Siegner [36]
Hemophilia B	FIX	ABE	Huh7 cells	Transfection with Lipofectamine 2000	Rong [37]
Sickle cell disease and β-thalassemia	HBG	CBE, ABE	HSPC	nucleofection	Antoniou [38]
β-thalassaemia	HBB	ABE	HSPC	nucleofection	Badat [39]
Sickle cell disease and β-thalassemia	HBG	CBE, ABE	HSPC	Electroporation	Ravi [40]
Sickle cell disease	HBB	ABE	HSPC	Electroporation	Newby [41]
β-thalassemia and sickle cell disease	HBG1, HBG2	ABE	HSPC	Electroporation	Mayuranathan [42]
β-thalassemia	BCL11A, HBG	ABE	HSPC	Electroporation	Liao [43]
β-thalassemia or sickle cell disease	HBG1/2	ABE	CD46/β-YAC mice	s.c. and i.v. injections	Li [44]
hemophilia B	F9	CBE	iPSC	electroporation	Hiramoto [45]
β-Thalassemia	HBB	ABE	HSPC	electroporation	Hardouin [46]
b-thalassemia and sickle cell disease	HBG1, HBG2	CBE	HSPC	electroporation	Han [47]
Spinal muscular atrophy	Smn2	ABE	patient-derived fibroblasts and SMA mouse	Lipofectamine LTX, intracerebroventricular (ICV) injections of AAV9	Alves [48]
Alzheimer's disease	APP	CBE	SH-SY5Y neuroblastoma	Transfection with Lipofectamine 2000	Guyon [49]
Spinal muscular atrophy	Smn2	ABE	Δ7SMA mice	Intracerebroventricular injections of AAV9	Arbab [50]
Alpha-1 antitrypsin deficiency	SERPINA1	ABE	iPSC	Nucleofection	Werder [51]
Mucopolysaccharidosis type I	Idua	ABE	murine model harboring the Idua-W392X mutation	Split AAV9, temporal vein injection	Su [52]
Hemochromatosis	Hfe	ABE	129-Hfetm.1.1Nca mice.	Split AAV9, tail vein injection	Rovai [53]

Disease	Gene	Base editor	*In vitro*/*in vivo*	Delivery method	Reference
Alpha-1 antitrypsin deficiency	Serpina1	ABE	PiZ-transgenic mice	tail vein injection of LNP	Packer [54]
Ganglioside-monosialic acid gangliosidosis	GLB1	ABE	Patient derived fibroblasts	Transfection with Viafect reagent	Leclerc [55]
Free sialic acid storage disorders	SLC17A5	ABE	human dermal fibroblasts	electroporation	Harb [56]
STAT3-hyper IgE syndrome	STAT3	ABE	patient fibroblasts and iPSCs	Electroporation	Eberherr [57]
Severe combined immune deficiency	CD3D	ABE	HSPC	electroporation	McAuley [58]
Familial dilated cardiomyopathy	Rbm20	ABE	iPSC, Homozygous (R636Q/R636Q) mice	Nucleofection, AAV9	Nishiyama [59]
Hypertrophic cardiomyopathy	Myh7	ABE	iPSC, humanized mouse	Nucleofection, injection into the inferior mediastinum	Chai [60]
Dilated cardiomyopathy	Rbm20	ABE	Rbm20 knock-in mouse	tail vein of AAV9	Grosch [61]
Retinitis pigmentosa	Pde6b	ABE	rd10 mouse model	Subretinal injections of Dual adenoassociated virus	Wu [62]
Usher syndrome type 1F	Pcdh15	ABE	humanized Pcdh15R245X mouse	dual AAV9	Peters [63]
Dominant-inherited deafness	Myo6	RNA base editing	Myo6C442Y/C442Y mice	Cochlea injection of AAV	Xiao [64]
Duchenne muscular dystrophy	Dmd	ABE	adult mice	AAV9	Xu [65]
Facioscapulohumeral muscular dystrophy	DUX4	ABE	immortalized myoblasts	Transfection with Lipofectamine 3000	Sikrová [66]
Recessive dystrophic epidermolysis bullosa	COL7A1	ABE	patient fibroblasts	Electroporation	Sheriff [67]
Recessive dystrophic epidermolysis bullosa	COL7A1	ABE	patient-derived fibroblasts	electroporation	Hong [68]
Cystic fibrosis	CFTR	ABE	patient-derived rectal organoids and bronchial epithelial cells	electroporation	Amistadi [69]

Table 1.
Selected examples of base editing on correcting pathogenic mutations.

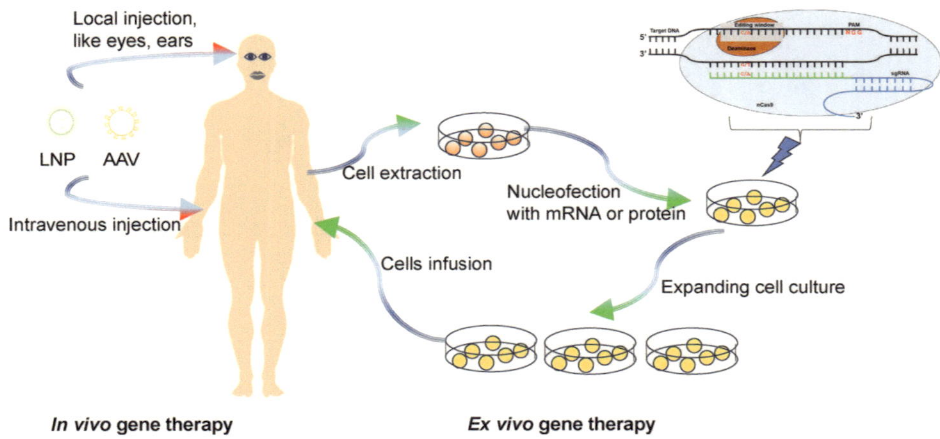

Figure 1.
In vivo *and* ex vivo *gene therapy based on base editing. For* ex vivo, *the nucleofection is the main method to deliver the editing system into the cells. For* in vivo, *AAV and LNP are the two main methods to deliver the base editing system into the target cells.*

Dozens of hematological disorders have been reported [72]. However, in terms of gene therapy, most studies have focused on sickle cell disease (SCD) and β-thalassemia [73, 74]. Both conditions are the result of mutations within the β-globin locus. Antoniou et al. utilized base editing to introduce mutations in the −200 region of the HBG promoter, leading to the activation of fetal hemoglobin [38]. This same strategy has been employed by other research groups [40, 42, 43, 47]. Li et al. implemented this strategy in healthy CD46/β-YAC mice carrying the human β-globin locus. They achieved editing through subcutaneous and intravenous injections, eliminating the need for myeloablation and hematopoietic stem cell transplantation [44]. Badat et al. used the ABE system to correct the HbE mutation to either wild-type (WT) or a normal variant hemoglobin (E26G). Ninety percent of primary human CD34+ cells were edited, and long-term repopulating hematopoietic stem cells (HSCs) were confirmed through xenotransplantation in NSG mice [39]. Editing the mutation in HBB represents a widespread strategy for treating SCD or β-thalassemia [41, 46].

Fanconi Anemia (FA) is a hematological disorder characterized by a wide range of severe symptoms. The DSB repair pathway is defective in FA, making it difficult to use HDR for mutation correction. Two groups have demonstrated that base editing can provide an effective method to correct mutations in the FANCA gene [35, 36]. Hemophilia is a congenital hemorrhagic disease caused by mutations in coagulation factor VIII (FVIII) or factor IX (FIX). Hiramoto et al. corrected a mutation ((c.947 T > C; I316T) in the F9 gene using CBE in induced pluripotent stem cells [45]. Rong et al. used ABE8e to correct G20519A and then measured the concentration and activity of FIX [37].

4.3 Neurodegenerative disease

Spinal muscular atrophy (SMA) is a devastating neuromuscular disease that often leads to infant mortality. It is caused by mutations in the SMN1 gene, with the most common mutation being the deletion of exon 7. The function of SMN1 can be partially compensated by SMN2, which differs from SMN1 primarily by a synonymous C·G-to-T·A transition in exon 7. This mutation results in insufficient expression. Arbab et al.

used base editing to correct this mutation in Δ7SMA mice. The mice achieved 87% conversions, and their motor function was enhanced [50]. This similar work has been replicated by another group [48].

Alzheimer's disease (AD) is a widely recognized condition that affects millions of people worldwide. Mutations in the APP gene are thought to play a role in the development of AD. Guyon et al. used ABE to convert the A673T mutation in SH-SY5Y neuroblastoma cells. Additionally, they identified a novel mutation, E674K, which appeared to further reduce Ab peptide accumulation. This finding may represent a new strategy for the treatment of Alzheimer's disease [49].

4.4 Inborn errors of metabolism

Inborn errors of metabolism (IEM) encompass over 1500 distinct diseases, primarily resulting from mutations in genes coding for proteins involved in intermediate metabolism. Nearly 25% of IEMs manifest during the neonatal period, emphasizing the importance of early intervention. Numerous gene therapy studies have focused on IEM using CRISPR/Cas9, with the delivery of the editing system as the primary challenge [75]. For base editing, most research in IEM has centered on delivering the editing system to target cells, often using induced pluripotent stem cells (iPSCs), fibroblasts, and mouse models to demonstrate the editing potential [51–56]. Curiously, most studies have utilized ABE rather than CBE for these corrections (**Table 1**). Additionally, genes such as ANGPTL3 and PCSK9 have been targeted to prevent atherosclerotic cardiovascular disease (ASCVD) [76, 77].

4.5 Other inherited diseases

Inborn errors of immunity (IEI) are disorders that affect the innate or adaptive immune system. STAT3-hyper IgE syndrome, a representative IEI, is characterized by destructive lung disease and various other symptoms. Eberherr et al. employed ABEs to correct the STAT3 p.R382W mutation in patient fibroblasts and iPSCs. The editing efficiency was approximately 29% in the patient fibroblasts, and the edited cells exhibited functional rescue of STAT3 signaling [57]. CD3δ severe combined immune deficiency (SCID), another life-threatening IEI, is caused by biallelic mutations in the autosomal CD3D gene. McAuley et al. utilized an ABE strategy to restore CD3δ in HSPCs, achieving an editing efficiency of 71.2% [58]. They also highlighted the advantages of base editing compared to standard CRISPR/Cas9 editing, indicating that base editing holds significant promise for future therapeutic applications [78].

Dystrophic epidermolysis bullosa (DEB) is a severe, rare disease characterized by skin blistering and epithelial fragility. It is caused by mutations in the COL7A1 gene. Sheriff et al. employed ABE technology to correct a recurrent COL7A1 nonsense mutation, c.5047 C > T (p.Arg1683Ter), in exon 54 of COL7A1 [67]. Hong et al. further demonstrated the potential of ABE to correct other mutations in the COL7A1 gene [68].

Cystic fibrosis (CF) is a genetic disease caused by mutations in the CF transmembrane conductance regulator (CFTR) gene. Various mutations of CFTR have been reported, including nonsense and splicing mutations, which have been corrected using base editing [69, 79, 80]. However, the most common mutation, F508del, exceeds the editing scope of base editing. Therefore, prime editing and homology-directed repair (HDR) are alternative choices for this mutation [81].

Duchenne muscular dystrophy (DMD) is a fatal genetic muscle disease that primarily affects males. Numerous research efforts are ongoing to develop effective therapies,

Identification code	Intervention / treatment	Phase	Sponsor	Delivery	Conditions	Target	Base editor
NCT05456880	BEAM-101	Phase 1/2	Beam Therapeutics Inc.	Electroporation, Ex vivo HSCs	Sickle Cell Disease and, β-Thalassemia	HBG1/2	ABE
NCT05885464	BEAM-201	Phase 1/2	Beam Therapeutics Inc.	Electroporation, Ex vivo T cells	T-ALL/T-LL, CD7 + AML	CD7, TRAC, PDCD1, CD52	CBE
NCT05398029	VERVE-101	Phase 1	Verve Therapeutics, Inc.	LNP, in vivo	HeFH	PCSK9	ABE
NCT06164730	VERVE-102	Phase 1	Verve Therapeutics, Inc.	GalNAc-LNP, in vivo	HeFH	PCSK9	ABE

Table 2.
Approved clinical trials using base editing.

including gene therapy based on CRISPR/Cas9 [82]. Xu et al. administered AAV9-iABE-NGA to adult mice, resulting in dystrophin restoration and functional improvement [65]. The off-target effect was low, and no significant toxicity was observed.

Numerous gene therapy studies based on base editing for other inherited diseases are currently underway [61, 64, 83, 84]. The proof-of-concept work has demonstrated the effectiveness and safety of base editing. The development of an effective delivery method *in situ* will further enhance the application of base editing.

5. Clinical trials using base editing approved by FDA

With the recent approval of the first gene editing product, we can expect to see a surge in the approval of additional gene editing products within the next decade. Currently, there are four products in clinical trials approved by FDA that utilize base editing techniques, produced by Beam therapeutics and Verve therapeutics (**Table 2**).

BEAM-101 and BEAM-201 are two products offered by Beam Therapeutics Inc. BEAM-101 is designed for the treatment of sickle cell disease and β-Thalassemia. It utilizes ABE to target the regulatory element of HBG, with the aim of activating the expression of fetal hemoglobin. The HSCs (hematopoietic stem cells) from patients are electroporated with mRNA encoding the base editor and sgRNA (single guide RNA). The edited cells are then infused back into the patients. BEAM-201, like BEAM-101, is also produced *ex vivo*. However, BEAM-201 is a CART (chimeric antigen receptor T cell) product designed to treat cancers. It utilizes CBE (cytosine base editor) to knock out CD7, TRAC, PDCD1, and CD52 genes [9].

VERVE-101 and VERVE-102 are developed by Verve Therapeutics, Inc. Both of them utilize an *in vivo* approach, which involves the use of LNP (lipid nanoparticle) to deliver the ABE (base editor) system targeting PCSK9 to the liver [77]. The difference between them lies in the modification of the LNP for VERVE-102 with galNAc. VERVE-101 is the first *in vivo* drug to utilize base editing, and it was selected by Nature Medicine as one of the 11 clinical trials that will shape medicine in 2024 [85].

6. Discussion and conclusion

Currently, more than 20 gene editing drugs are in clinical trials, and with the first gene editing drugs approved, we can expect a growing number of gene editing drugs to gain approval in the future. These drugs are being developed not only for inherited diseases but also for cancers, infection diseases, and even HIV by biotech companies and large pharma.

6.1 Gene therapy *in utero*

Recently, most of these therapies have been directed toward editing somatic cells, through either *ex vivo* or *in vivo* methods. Germline editing is currently only permitted for basic research purposes, and it remains forbidden in clinical settings. *In utero* gene therapy (IUGT) has gained attention, and many researches have been conducted on animals [86]. There are several advantages for IUGT, such as small fetal size, tolerogenic immune system, and prevalent progenitor cells [87]. Transgenic rodent models have demonstrated the long-term safety of IUGT, but several issues remain to be addressed. One concern is the potential risk to the maternal bystander, while another is the risk of germline mutations. Considering the potential risk, disease selection for IUGT is important. The most suitable scene for IUGT is the lethal and morbid perinatal diseases. For different diseases, the best therapy time is before the onset of pathology. Optimization of the delivery methods focused on the fetus will enhance the clinical translation of IUGT.

6.2 Efficiency and precision of base editing

Currently, most base editors have an editing window, during which any C or A can be edited. This characteristic introduces bystander editing, which needs to be considered when using base editing to correct pathogenic mutations. If the aim is to disrupt a gene, bystander editing may be less significant. In the future, narrowing the editing window may help precisely correct pathogenic mutations. CRISPR/Cas requires a PAM sequence to bind the target site. Currently, there are various Cas proteins, including Cas9-SpRY, which do not have the PAM limitation [88]. However, SpRY has the disadvantage of targeting itself when the plasmid is used. Ideally, a Cas protein should have PAM limitations but be able to target different PAM sequences for NA, NT, NG, or NC. For *ex vivo* therapy, electroporation is commonly used. Electroporation of proteins may help reduce potential off-target effects. In the future, it is important for the engineered editor to easily purify its protein and ensure protein stability.

6.3 Delivery method of the base editing system

Gene therapy can be classified into *ex vivo* and *in vivo* methods. In the *ex vivo* approach, target cells like HSPCs and immune cells are isolated, edited, and then infused into the patient. The procedure for *ex vivo* therapy is relatively mature; however, it is associated with high costs due to its complexity. In contrast, gene therapy *in vivo* has the potential to reduce costs. The main obstacle for *in vivo* therapy is delivery, and viral delivery is the most efficient method currently available. Nonviral delivery systems represent a promising future direction for *in vivo* gene therapy. Among nonviral delivery systems, LNPs show advantages for liver delivery, but further development of specific delivery methods is needed for other tissues [7].

6.4 Price of gene editing products

Gene therapy drugs offer great hope to patients, but their high cost makes it difficult to apply them to all patients, which may be unfair to those in poverty. Additionally, the high price may result in the delisting of drugs, as seen with Glybera, a gene therapy product approved in 2012 but withdrawn in 2017 due to its exorbitant cost. The complexity of the manufacturing process for these drugs contributes to their high production costs. To address this issue, more reasonable payment methods and the development of off-the-shelf gene therapy drugs may offer potential solutions.

In conclusion, base editing has demonstrated advantages over traditional editing methods based on DSB. Numerous clinical trials using base editing are currently underway. With the obstacles discussed above addressed, we believe that gene therapy has the potential to edit diseases out.

Acknowledgements

This work was supported by the National Key Research and Development Program of China (2021YFA0804702) and the National Natural Science Foundation of China (82302497). We thank members of Huang's lab for helpful discussions.

Conflict of interest

The authors declare no conflict of interest.

Author details

Guanglei Li[1*], Yu Zhang[1] and Xingxu Huang[1,2*]

1 Gene Editing Center, School of Life Science and Technology, ShanghaiTech University, Shanghai, China

2 The Key Laboratory of Pancreatic Diseases of Zhejiang Province, the First Affiliated Hospital, Zhejiang University School of Medicine, Hangzhou, China

*Address all correspondence to: huangxx@shanghaitech.edu.cn and ligl@shanghaitech.edu.cn

References

[1] Wong C. UK first to approve CRISPR treatment for diseases: What you need to know. Nature. 2023;**623**:676-677

[2] Jinek M et al. A programmable dual-RNA-guided DNA endonuclease in adaptive bacterial immunity. Science. 2012;**337**:816-821

[3] Frangoul H et al. CRISPR-Cas9 gene editing for sickle cell disease and beta-thalassemia. The New England Journal of Medicine. 2021;**384**:252-260

[4] Ledford H. CRISPR 2.0: A new wave of gene editors heads for clinical trials. Nature. 2023;**624**:234-235

[5] Braga LAM, Conte Filho CG, Mota FB. Future of genetic therapies for rare genetic diseases: What to expect for the next 15 years? Therapeutic Advances in Rare Disease. 2022;**3**:26330040221100840

[6] Roth TL, Marson A. Genetic disease and therapy. Annual Review of Pathology. 2021;**16**:145-166

[7] Huang C, Li Q, Li J. Site-specific genome editing in treatment of inherited diseases: Possibility, progress, and perspectives. Medicine Review (Berl). 2022;**2**:471-500

[8] Liu X et al. Advances in CRISPR/Cas gene therapy for inborn errors of immunity. Frontiers in Immunology. 2023;**14**:1111777

[9] Diorio C et al. Cytosine base editing enables quadruple-edited allogeneic CART cells for T-ALL. Blood. 2022;**140**:619-629

[10] Altae-Tran H et al. Uncovering the functional diversity of rare CRISPR-Cas systems with deep terascale clustering. Science. 2023;**382**:eadi1910

[11] Bendixen L, Jensen TI, Bak RO. CRISPR-Cas-mediated transcriptional modulation: The therapeutic promises of CRISPRa and CRISPRi. Molecular Therapy. 2023;**31**:1920-1937

[12] Goell JH, Hilton IB. CRISPR/Cas-based Epigenome editing: Advances, applications, and clinical utility. Trends in Biotechnology. 2021;**39**:678-691

[13] Haapaniemi E, Botla S, Persson J, Schmierer B, Taipale J. CRISPR-Cas9 genome editing induces a p53-mediated DNA damage response. Nature Medicine. 2018;**24**:927-930

[14] Tsuchida CA et al. Mitigation of chromosome loss in clinical CRISPR-Cas9-engineered T cells. Cell. 2023;**186**:4567-4582 e4520

[15] Dang L et al. Comparison of gene disruption induced by cytosine base editing-mediated iSTOP with CRISPR/Cas9-mediated frameshift. Cell Proliferation. 2020;**53**:e12820

[16] Amendola M, Brusson M, Miccio A. CRISPRthripsis: The risk of CRISPR/Cas9-induced Chromothripsis in gene therapy. Stem Cells Translational Medicine. 2022;**11**:1003-1009

[17] Komor AC, Kim YB, Packer MS, Zuris JA, Liu DR. Programmable editing of a target base in genomic DNA without double-stranded DNA cleavage. Nature. 2016;**533**:420-424

[18] Koblan LW et al. Efficient C*G-to-G*C base editors developed using CRISPRi screens, target-library analysis, and machine learning. Nature Biotechnology. 2021;**39**:1414-1425

[19] Liu Y et al. A Cas-embedding strategy for minimizing off-target effects of DNA base editors. Nature Communications. 2020;**11**:6073

[20] Wang L et al. Eliminating base-editor-induced genome-wide and transcriptome-wide off-target mutations. Nature Cell Biology. 2021;**23**:552-563

[21] Wang X et al. Efficient gene silencing by Adenine Base editor-mediated start codon mutation. Molecular Therapy. 2020;**28**:431-440

[22] Porto EM, Komor AC. In the business of base editors: Evolution from bench to bedside. PLoS Biology. 2023;**21**:e3002071

[23] Kar B, Castillo SR, Sabharwal A, Clark KJ, Ekker SC. Mitochondrial Base editing: Recent advances towards therapeutic opportunities. International Journal of Molecular Sciences. 2023;**24**:5798

[24] Pfeiffer LS, Stafforst T. Precision RNA base editing with engineered and endogenous effectors. Nature Biotechnology. 2023;**41**:1526-1542

[25] Anzalone AV et al. Search-and-replace genome editing without double-strand breaks or donor DNA. Nature. 2019;**576**:149-157

[26] Doman JL et al. Phage-assisted evolution and protein engineering yield compact, efficient prime editors. Cell. 2023;**186**:3983-4002 e3926

[27] Godbout K, Tremblay JP. Prime editing for human gene therapy: Where are we now? Cell. 2023;**12**:536

[28] Li G et al. Gene editing and its applications in biomedicine. Science China. Life Sciences. 2022;**65**:660-700

[29] Kim K et al. Highly efficient RNA-guided base editing in mouse embryos. Nature Biotechnology. 2017;**35**:435-437

[30] Ryu SM et al. Adenine base editing in mouse embryos and an adult mouse model of Duchenne muscular dystrophy. Nature Biotechnology. 2018;**36**:536-539

[31] Rosello M et al. Disease modeling by efficient genome editing using a near PAM-less base editor *in vivo*. Nature Communications. 2022;**13**:3435

[32] Zheng S et al. Efficient PAM-Less Base editing for zebrafish modeling of human genetic disease with zSpRY-ABE8e. Journal of Visualized Experiments. 2023;**192**

[33] Wang F et al. Generation of a Hutchinson-Gilford progeria syndrome monkey model by base editing. Protein & Cell. 2020;**11**:809-824

[34] Zheng X et al. Generation of inactivated IL2RG and RAG1 monkeys with severe combined immunodeficiency using base editing. Signal Transduction and Targeted Therapy. 2023;**8**:327

[35] Sipe CJ et al. Correction of Fanconi anemia mutations using digital genome engineering. International Journal of Molecular Sciences. 2022;**23**:8416

[36] Siegner SM et al. Adenine base editing efficiently restores the function of Fanconi anemia hematopoietic stem and progenitor cells. Nature Communications. 2022;**13**:6900

[37] Rong L, Chen D, Huang X, Sun L. Delivery of Cas9-guided ABE8e into stem cells using poly(l-lysine) polypeptides for correction of the hemophilia-associated FIX missense mutation. Biochemical and Biophysical Research Communications. 2022;**628**:49-56

[38] Antoniou P et al. Base-editing-mediated dissection of a gamma-globin cis-regulatory element for the therapeutic reactivation of fetal

hemoglobin expression. Nature Communications. 2022;**13**:6618

[39] Badat M et al. Direct correction of haemoglobin E beta-thalassaemia using base editors. Nature Communications. 2023;**14**:2238

[40] Ravi NS et al. Identification of novel HPFH-like mutations by CRISPR base editing that elevate the expression of fetal hemoglobin. eLife. 2022;**11**:e65421

[41] Newby GA et al. Base editing of haematopoietic stem cells rescues sickle cell disease in mice. Nature. 2021;**595**:295-302

[42] Mayuranathan T et al. Potent and uniform fetal hemoglobin induction via base editing. Nature Genetics. 2023;**55**:1210-1220

[43] Liao J et al. Therapeutic adenine base editing of human hematopoietic stem cells. Nature Communications. 2023;**14**:207

[44] Li C et al. *In vivo* base editing by a single i.v. vector injection for treatment of hemoglobinopathies. JCI Insight. 2022;**7**:e162939

[45] Hiramoto T et al. PAM-flexible Cas9-mediated base editing of a hemophilia B mutation in induced pluripotent stem cells. Communications Medicine (Lond). 2023;**3**:56

[46] Hardouin G et al. Adenine base editor-mediated correction of the common and severe IVS1-110 (G>a) beta-thalassemia mutation. Blood. 2023;**141**:1169-1179

[47] Han W et al. Base editing of the HBG promoter induces potent fetal hemoglobin expression with no detectable off-target mutations in human HSCs. Cell Stem Cell. 2023;**30**:1624-1639 e1628

[48] Alves CRR et al. Optimization of base editors for the functional correction of SMN2 as a treatment for spinal muscular atrophy. Nature Biomedical Engineering. 2023

[49] Guyon A et al. Base editing strategy for insertion of the A673T mutation in the APP gene to prevent the development of AD *in vitro*. Molecular Therapy - Nucleic Acids. 2021;**24**:253-263

[50] Arbab M et al. Base editing rescue of spinal muscular atrophy in cells and in mice. Science. 2023;**380**:eadg6518

[51] Werder RB et al. Adenine base editing reduces misfolded protein accumulation and toxicity in alpha-1 antitrypsin deficient patient iPSC-hepatocytes. Molecular Therapy. 2021;**29**:3219-3229

[52] Su J et al. In vivo adenine base editing corrects newborn murine model of hurler syndrome. Molecular Biomedicine. 2023;**4**:6

[53] Rovai A et al. In vivo adenine base editing reverts C282Y and improves iron metabolism in hemochromatosis mice. Nature Communications. 2022;**13**:5215

[54] Packer MS et al. Evaluation of cytosine base editing and adenine base editing as a potential treatment for alpha-1 antitrypsin deficiency. Molecular Therapy. 2022;**30**:1396-1406

[55] Leclerc D et al. Gene editing corrects In vitro a G > a GLB1 transition from a GM1 Gangliosidosis patient. CRISPR Journal. 2023;**6**:17-31

[56] Harb JF et al. Base editing corrects the common Salla disease SLC17A5 c.115C>T variant. Molecular Therapy - Nucleic Acids. 2023;**34**:102022

[57] Eberherr AC et al. Rescue of STAT3 function in hyper-IgE syndrome using

Adenine Base editing. CRISPR Journal. 2021;**4**:178-190

[58] McAuley GE et al. Human T cell generation is restored in CD3delta severe combined immunodeficiency through adenine base editing. Cell. 2023;**186**:1398-1416 e1323

[59] Nishiyama T et al. Precise genomic editing of pathogenic mutations in RBM20 rescues dilated cardiomyopathy. Science Translational Medicine. 2022;**14**:eade1633

[60] Chai AC et al. Base editing correction of hypertrophic cardiomyopathy in human cardiomyocytes and humanized mice. Nature Medicine. 2023;**29**:401-411

[61] Grosch M et al. Striated muscle-specific base editing enables correction of mutations causing dilated cardiomyopathy. Nature Communications. 2023;**14**:3714

[62] Wu Y et al. AAV-mediated base-editing therapy ameliorates the disease phenotypes in a mouse model of retinitis pigmentosa. Nature Communications. 2023;**14**:4923

[63] Peters CW et al. Rescue of hearing by adenine base editing in a humanized mouse model of usher syndrome type 1F. Molecular Therapy. 2023;**31**:2439-2453

[64] Xiao Q et al. Rescue of autosomal dominant hearing loss by in vivo delivery of mini dCas13X-derived RNA base editor. Science Translational Medicine. 2022;**14**:eabn0449

[65] Xu L et al. Efficient precise in vivo base editing in adult dystrophic mice. Nature Communications. 2021;**12**:3719

[66] Sikrova D et al. Adenine base editing of the DUX4 polyadenylation signal for targeted genetic therapy in facioscapulohumeral muscular dystrophy. Molecular Therapy - Nucleic Acids. 2021;**25**:342-354

[67] Sheriff A et al. ABE8e adenine base editor precisely and efficiently corrects a recurrent COL7A1 nonsense mutation. Scientific Reports. 2022;**12**:19643

[68] Hong SA et al. Therapeutic base editing and prime editing of COL7A1 mutations in recessive dystrophic epidermolysis bullosa. Molecular Therapy. 2022;**30**:2664-2679

[69] Amistadi S et al. Functional restoration of a CFTR splicing mutation through RNA delivery of CRISPR adenine base editor. Molecular Therapy. 2023;**31**:1647-1660

[70] Li Y, Glass Z, Huang M, Chen ZY, Xu Q. Ex vivo cell-based CRISPR/Cas9 genome editing for therapeutic applications. Biomaterials. 2020;**234**:119711

[71] Madigan V, Zhang F, Dahlman JE. Drug delivery systems for CRISPR-based genome editors. Nature Reviews. Drug Discovery. 2023;**22**:875-894

[72] Echeverry G, Dalton A. Hematologic disorders. Anesthesiology Clinics. 2018;**36**:553-565

[73] Park SH, Bao G. CRISPR/Cas9 gene editing for curing sickle cell disease. Transfusion and Apheresis Science. 2021;**60**:103060

[74] Ma L, Yang S, Peng Q, Zhang J, Zhang J. CRISPR/Cas9-based gene-editing technology for sickle cell disease. Gene. 2023;**874**:147480

[75] Leal AF et al. The landscape of CRISPR/Cas9 for inborn errors of metabolism. Molecular Genetics and Metabolism. 2023;**138**:106968

[76] Zuo Y et al. Liver-specific in vivo base editing of Angptl3 via AAV delivery efficiently lowers blood lipid levels in mice. Cell & Bioscience. 2023;**13**:109

[77] Lee RG et al. Efficacy and safety of an investigational single-course CRISPR Base-editing therapy targeting PCSK9 in nonhuman primate and mouse models. Circulation. 2023;**147**:242-253

[78] Malech HL, Notarangelo LD. Gene therapy for inborn errors of immunity: Base editing comes into play. Cell. 2023;**186**:1302-1304

[79] Geurts MH et al. CRISPR-based adenine editors correct nonsense mutations in a cystic fibrosis organoid biobank. Cell Stem Cell. 2020;**26**:503-510 e507

[80] Graham C, Hart S. CRISPR/Cas9 gene editing therapies for cystic fibrosis. Expert Opinion on Biological Therapy. 2021;**21**:767-780

[81] Geurts MH et al. Evaluating CRISPR-based prime editing for cancer modeling and CFTR repair in organoids. Life Science Alliance. 2021;**4**:e202000940

[82] Verhaart IEC, Aartsma-Rus A. Therapeutic developments for Duchenne muscular dystrophy. Nature Reviews. Neurology. 2019;**15**:373-386

[83] Koblan LW et al. In vivo base editing rescues Hutchinson-Gilford progeria syndrome in mice. Nature. 2021;**589**:608-614

[84] Kaukonen M, McClements ME, MacLaren RE. CRISPR DNA Base editing strategies for treating retinitis Pigmentosa caused by mutations in rhodopsin. Genes (Basel). 2022;**13**:1327

[85] Arnold C, Webster P. 11 clinical trials that will shape medicine in 2024. Nature Medicine. 2023;**29**:2964-2968

[86] Mattar CNZ, Chan JKY, Choolani M. Gene modification therapies for hereditary diseases in the fetus. Prenatal Diagnosis. 2023;**43**:674-686

[87] Bose SK, Menon P, Peranteau WH. InUtero gene therapy: Progress and challenges. Trends in Molecular Medicine. 2021;**27**:728-730

[88] Anzalone AV, Koblan LW, Liu DR. Genome editing with CRISPR-Cas nucleases, base editors, transposases and prime editors. Nature Biotechnology. 2020;**38**:824-844

Section 3

Pregnancy and Parturition

Chapter 5

Perspective Chapter: A New View of the Endocrinology of Pregnancy and Parturition – Lessons from the Literature

Christopher Edwards

Abstract

Human and sheep parturition are more akin than currently recognised. In both glucocorticoids are key. The difference being mechanisms controlling glucocorticoid levels. Sheep have low cortisol during pregnancy which rise at term: humans control local glucocorticoid levels via the fetal adrenal and DHEA-sulphate. This increases 11β-HSD2 expression protecting the fetus from maternal cortisol by converting this to cortisone. During pregnancy DHEA inhibits placental and fetal membrane 11β-HSD1 expression. This plus hexose-6-phosphate dehydrogenase inhibition decreases 11β-HSD1 oxido-reductase/increases dehydrogenase action converting cortisol to cortisone via transcription factor C/EBPβ. This has a key role in progesterone synthesis blocking COX-2 transcription and prostaglandin synthesis. DHEA-induced protection reverses at parturition onset via pro-inflammatory cytokines increasing C/EBPα. This lowers progesterone and enhances *local* glucocorticoid production stimulating prostaglandins, oxytocin receptor production and cervical ripening. Lowered progesterone and increased TNFα/IL-1β markedly increase myometrial purinergic receptors promoting calcium entry, contraction and hence parturition.

Keywords: cortisol, DHEA, 11β-HSD2 and 11β-HSD1, fetal adrenal, hexose-6-phosphate dehydrogenase

1. Introduction

It is clear that the factors controlling parturition are complex and poorly understood. Several studies suggest that the fetus is in control of the timing of the onset of labour. In sheep and cows activation at term of the fetal hypothalamic-pituitary-adrenal axis (HPA) plays a major role. Cortisol directly upregulates the placental enzyme 17α-hydroxylase/17,20-lyase (CYP17) that catalyses the conversion of pregnenolone to 17-OH-Pregnenolone and then to Dehydroepiandrosterone (DHEA). DHEA can then be further metabolised to oestrogen (oestriol from DHEA-S and oestradiol and oestrone from testosterone and androstenedione). It is suggested that the switch in the progesterone/oestrogen ratio then provides the stimulus for uterine prostaglandin

IntechOpen

production and hence labour. Oestrogens are thought to increase prostaglandin and oxytocin receptors, and upregulate myosin light chain kinase that enhances myometrial contraction. This pathway however was thought not apply in humans as human placentae were said to lack the CYP17 enzyme. More recent work has suggested that this is not true. Escobar and colleagues have shown that the human placenta does express CYP17 and can generate androgens *de novo* in addition to those produced by the fetal adrenal [1]. Others have confirmed this [2].

Many authors have suggested that the human placenta is an 'incomplete steroidogenic organ' and that oestrogen synthesis by the placenta requires the C19 oestrogen precursor DHEA-S that is produced by the fetal adrenal. It is suggested that DHEA-S is metabolised by the fetal liver to 16-hydroxy DHEA-S that then travels to the placenta where it is metabolised to oestrogen.

This chapter focuses on a possible novel role for DHEA and DHEA-S and suggests that these steroids are not simply produced as oestrogen precursors but regulate the local intracellular concentrations of glucocorticoid in critical sites and at key times in pregnancy.

2. The human fetal adrenal

Ishimoto and Jaffe have written an excellent review of the development and function of the human fetal adrenal cortex [3]. They summarise the evidence that there is early cortisol secretion by the fetal adrenal followed by suppressed synthesis until late gestation. Studies in infants with 21-hydroxylase deficiency (who cannot synthesise adequate amounts of cortisol) show that they have a compensatory rise in ACTH that produces fetal adrenal hyperplasia and associated adrenal androgen production (DHEA-S). Maternal administration of dexamethasone that crosses the placenta reverses the virilisation of the external genitalia if given before the 7th or 8th week of gestation.

One of the major problems in studying fetal adrenal function has been the major species differences so that rodents cannot be used as a model [4].

Goto [5] has suggested that in the normal fetus early cortisol synthesis has a negative feedback effect on ACTH secretion which lowers ACTH-induced androgen secretion and safeguards normal female sexual development.

Despite the lack of cortisol secretion by the fetal adrenal for the majority of pregnancy adrenal size suggests that this is not its function. By 20 weeks gestation the fetal adrenal is as big as the fetal kidney and by 30 weeks it achieves a relative size that is 10–20 fold that of the adult adrenal gland. This rapid growth is almost entirely due to enlargement of the fetal zone that is responsible for the production of DHEA-S. By 16–20 weeks the fetal zone (FZ) dominates the gland.

A key enzyme in the conversion of pregenolone to progesterone, 17-OH-pregenolone to 17-OH-progesterone and DHEA to androstenedione is HSD3B2 (3β-HSD) (**Figure 1**). Levels of this enzyme peak in the adrenal at 8–9 weeks post-conception in keeping with the early peak of cortisol synthesis. With the exception of the first trimester HSD3B2 is not expressed in the fetal zone (i.e., oestrogens and androgens would need to be produced peripherally). After 23 weeks gestation HSD3B2 is detectable in both the Definitive Zone and the Transitional Zone. Interestingly HSD3B2 is readily expressed *in vitro* when stimulated by ACTH suggesting that ACTH is suppressed.

The key steroid produced by the fetal adrenal is DHEA-S. The localization of the DHEA sulfotransferase (SULT2A1) is thus of importance. It is found in the Fetal Zone

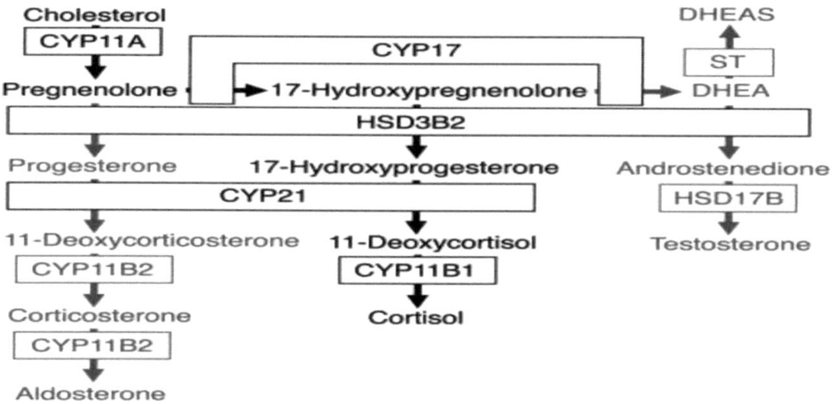

Figure 1.
Adrenal steroidogenesis.

and the Transitional Zone but not the Definitive Zone (DZ). The DZ is the likely site of aldosterone synthesis late in gestation (i.e., becomes the zona glomerulosa) and the TZ cortisol, again late in gestation (becomes zona fasciculata).

DHEA-S production starts at about 8–10 weeks gestation and progressively increases, so at term the fetal adrenals produce about 200 mg per day.

Beverley Murphy reviewed the data on cord blood levels of cortisol during human gestation [6]. At about 15 weeks the level was 8 ng/ml (22 nmol/l) falling to 4 at 17–20 weeks. It then rose to 20 ng/ml at 35–37 weeks and 45 ng/ml at term. The cortisol rise appeared to be steepest immediately before the onset of labour and could not be attributed to the stress of labour.

All this raises intriguing questions—why does the fetal adrenal produce such large quantities of DHEA-S and why is it produced as a sulfate rather than as DHEA? Given the amount produced and its time course it would seem unlikely that it is simply a source of oestrogen. The hypothesis put forward here suggests an alternative role.

3. The importance of cortisol metabolism in controlling local glucocorticoid levels

Our previous studies indicated the critical importance of local metabolism of cortisol and its inter-conversion with the inactive steroid cortisone [7]. Failure to convert cortisol to cortisone results in the syndrome of apparent mineralocorticoid excess where the mineralocorticoid receptor is exposed to cortisol which produces marked sodium retention, hypertension and hypokalemia [8]. The enzyme responsible for this is 11β-Hydroxysteroid Dehydrogenase type 2 (11β-HSD2). In contrast to this unidirectional NAD-dependent enzyme the one responsible for the inter-conversion of cortisol and cortisone is 11β-HSD1.

3.1 Hexose-6-phosphate dehydrogenase and its interaction with DHEA

Elegant studies by Paul Stewart et al. showed that 11β-HSD1 lives in the endoplasmic reticulum and has a very close relationship with hexose-6-phosphate dehydrogenase (H6PDH) [9]. H6PDH converts NADP to NADPH. This cofactor is then used by

11β-HSD1 to drive the enzyme as an oxido-reductase converting cortisone to cortisol. In the absence of H6PDH 11β-HSD1 acts as a dehydrogenase and converts cortisol to cortisone. This role was specific for H6PDH. Glucose-6-phosphate dehydrogenase was not effective.

3.2 Placental hexose-6-phosphate dehydrogenase

If the mechanism controlling the activity of 11β-HSD1 in the placenta is similar to that in, for example, adipose tissue then the presence of H6PDH to generate NADPH would be key. This enzyme, H6PDH, is found in high concentration in the human placenta, is located in the luminal space of the placental microsomes and generates NADPH on the inner side of the endoplasmic reticulum membrane [10].

3.3 The control of the expression and the set point of 11β-HSD1

Given the key role of H6PDH,what controls H6PDH?. This paper suggests that this could relate to DHEA-S. As in the fetus DHEA-S is the most abundant steroid produced by the adult adrenal cortex (20 times the secretion rate of cortisol). DHEA-S is produced by catarrhine primates—the Old World monkeys and the apes. Rodents do not produce large amounts of DHEA and this needs to be remembered when looking at animal models.

Marks and Banks [11] found that DHEA and pregnenolone were non-competitive inhibitors of G6PD. A concentration of DHEA of 1×10^{-5} M produced 82% inhibition of G6PD in a crude preparation of human liver.

The specificity of this reaction was further examined by Raineri and Levy [12]. They found that a 17-keto group was required for inhibition. Polar substitution at C-3 decreased the inhibitory activity but this was least with a β-OH (DHEA is 3β-hydroxy-5β-androstan-17-one).

Ziboh et al. looked at DHEA inhibition of lipid synthesis and G6PD in rat skin [13]. They found that synthesis from acetate was inhibited 44–56% by 2.5×10^{-4} M DHEA. This could be reversed by adding NADPH. They made the important point that the plasma concentrations of DHEA are between 5×10^{-9} and 2×10^{-8} M (i.e., 2 to 3 orders of magnitude below the concentrations required to be effective in inhibiting lipid synthesis *in vitro*).

Two papers have looked at the possibility that 11β-HSD1 is regulated by the pentose pathway flux; the first of these was by McCormick and colleagues [14]. They suggested that despite the fact that 11β-HSD1 is located in the endoplasmic reticulum that it is interconnected with the cytosolic pentose phosphate pathway, the primary producer of reduced pyridine nucleotides. They looked at this in intact rodent adipocytes. Inhibitors of NADPH generation via the pentose pathway (DHEA and norepinephrine) inhibited 11β-HSD1 oxido-reductase activity while decreasing cellular NADPH content. These compounds also augmented the reverse or dehydrogenase reaction (100 μM DHEA produced 187% increase in dehydrogenase activity). They then tested the compounds against isolated intact microsomes containing the tandem 11β-HSD1 and H6PDH unit. In this situation the inhibitors were not effective.

One of the surprising things about these results is that it is normally thought that the microsomal membrane is poorly permeable to pyridine nucleotides (i.e., that NADPH produced in the cytoplasm would not cross to enable 11β-HSD1 to act as a reductase). They incubated pyridine nucleotides with liver microsomes and found that NADPH very significantly increased 11β-HSD1 reductase activity. Interestingly NADP had no significant effect.

Another fascinating possibility needs consideration. It has been found that the sedoheptulose kinase CARKL directs macrophage polarisation through control of glucose metabolism [15]. Inhibition of CARKL by LPS, for example, results in a type 1 pro-inflammatory macrophage. This results in inhibition of the production of Fructose-6-phosphate (F-6-P) which normally contributes to glycolysis. Senesi et al. [16] showed that F-6-P contributes to glucocorticoid activation in the ER. It is transported into the ER and then isomerised to G-6-P which acts as the substrate for H6PDH, produces NADPH and activates the reductase action of 11β-HSD1.

A second key paper gives a different perspective. McNelis and colleagues [17] showed that DHEA had an anti-glucocorticoid action on human pre-adipocytes. Glucocorticoids stimulate human pre-adipocyte proliferation, differentiation and glucose uptake. Key to this is the local conversion of cortisone to cortisol by 11β-HSD1. The authors specifically looked at human cells given the relative lack of production of DHEA by rodent adrenals. They found that DHEA but not DHEA-S significantly inhibited pre-adipocyte proliferation and that this was due to reduced 11β-HSD1 oxido-reductase activity. This effect was significant at 1 μM DHEA and highly significant at 10 and 25 μM. At the highest dose (25 μM) there was a concurrent increase in dehydrogenase activity. DHEA also significantly increased insulin-independent glucose uptake.

The critically important part of the paper was the discovery that DHEA treatment inhibited the *expression* of both H6PDH and 11β-HSD1 in a dose-dependent manner. It was unclear whether this was by direct regulation of gene expression or via a regulator.

The question is whether these two mechanisms (i.e., non-competitive inhibition of G6PD and decreased expression of H6PDH) are both relevant and, possibly, synergistic. Thus could DHEA inhibit the pentose pathway in the cytoplasm and thus reduce the availability of F-6-P for transfer into the ER? If DHEA can also inhibit the expression of H6PDH then this would further reduce the intra-ER level of NADPH and hence decrease 11β-HSD1 conversion of cortisone to cortisol and, with high levels of DHEA, enhance conversion of cortisol to cortisone. Given the likelihood that the sulfatase in the endoplasmic reticulum is active the basic model could be as follows:

1. DHEA-S enters the cell via a specific transporter (OATP-D) and then crosses into the smooth endoplasmic reticulum (ER) where it is converted to DHEA by the sulfatase enzyme.

2. DHEA then diffuses out into the cytoplasm and inhibits the pentose phosphate pathway by direct binding to G6PD. This then reduces the amount of F-6-P that is available to cross the ER membrane to be isomerized to G-6-P and thus limits the production of NADPH and hence drives dehydrogenase activity of 11β-HSD1 in the ER.

3. DHEA also inhibits the expression of H6PDH. This then also lowers the production of NADPH within the ER.

The Odermatt group reported [18, 19] reduced 11β-HSD1 expression following treatment with DHEA in 3T3-L1 adipocytes and in liver and adipose tissue of treated mice. They found that the CCAAT/enhancer binding protein (C/EBP) family of transcription factors (C/EBP-α, C/EBP-β, and C/EBP-δ) were involved in the DHEA-mediated effect on 11β-HSD1 expression. C/EBP-α and C/EBP-β were shown to bind directly to the 11β-HSD1 promoter and acted in concert to regulate gene expression in liver cells and adipocytes.

C/EBP-α, a potent activator of 11β-HSD1 gene transcription, was downregulated by DHEA treatment in the liver and adipose tissue of the treated mice in comparison to C/EBP-β and C/EBP-δ which were either elevated or unchanged. Thus a low C/EBP-α/C/EBP-β ratio was associated with decreased 11β-HSD1 transcription.

They then went on to investigate whether C/EBP might also modulate HSD11B2 gene expression. They found that DHEA up-regulated the expression and activity of 11β-HSD2 in rat renal cortical collecting duct cells. The main effect appeared to be via activation of gene transcription. In comparison to 11β-HSD1 11β-HSD2 was more stimulated by C/EBP-β and a low C/EBP-α/C/EBP-β ratio was linked to increased expression of 11β-HSD2. They suggested that DHEA modulated the expression of 11β-HSD2 in a phosphatidylinositol-3-kinase/AKT-dependent manner by increasing C/EBP. In the context of this paper and parturition it is of interest that they investigated whether metabolism of DHEA to oestrogen or androgen was required for the DHEA effect on 11β-HSD2. Neither the oestrogen receptor antagonist tamoxifen nor the androgen receptor antagonist flutamide affected the DHEA-mediated upregulation of 11β-HSD2.

4. Transcription factors and 11β-HSD1/2 transcription

No such studies have been performed in relation to DHEA and the placental 11β-HSD2. It would of major physiological and potentially pathological interest if DHEA enhanced the transcription of placental 11β-HSD2 via this mechanism. DHEA did not affect 11β-HSD2 expression in Caco-2 and SW620 colon cells, suggesting tissue-specific regulation [19].

Chapman has reviewed the role and regulation of 11β-HSD1 in inflammation [20]. The enzyme is transcribed from three promoters, P1, P2 and P3. Liver and brain transcription is mainly from P2 and, as described above, is dependent on C/EBP as is the acute phase response. Mice lacking C/EBPα cannot mount such a response and die from neonatal hypoglycaemia. Pro-inflammatory cytokines interact with this mechanism. TNF-α increases the ratio of C/EBPα/C/EBPβ and hence enhances the transcription of 11β-HSD1. It is suggested that TNF promotes C/EBPβ phosphorylation and thus its export from the nucleus.

Working in a very different area Hardy et al. studied fibroblasts from patients with either rheumatoid arthritis or osteoarthritis [21]. These contained 11β-HSD1. Expression and activity of the enzyme increased following treatment with either IL-1β or TNFα (bone marrow: 37 and 8-fold respectively compared to vehicle; dermal fibroblasts 14-fold and 4-fold; synovial fibroblasts 31-fold and 7-fold; all $p < 0.001$ compared to vehicle). If this was happening in the amnion and 11β-HSD1 was acting as an oxido-reductase this could markedly increase the intracellular glucocorticoid levels.

This effect may well depend on the cell type used. Thus Ahasan et al. [22] have reported that TNFα and IL-1β increase the expression and activity of 11β-HSD1 in mesenchymal stromal cell types and tissues. This was different to what had been found in hepatocytes and blood cells. In addition combined treatment of glucocorticoids and pro-inflammatory cytokines *synergistically* increased the expression of 11β-HSD1 in synovial fibroblasts and myocytes. Using mouse embryonic fibroblasts they explored the cellular signalling pathways involved. They showed that the increased expression occurred via the proximal promoter and depended on NF-κB signalling. Targeted disruption of the RelA domain in the N-terminus of NF-κB prevented TNF and IL-1β enhancement of 11β-HSD1. Expression by glucocorticoids could be reproduced by specific inhibitors of p38 MAPK. Mesenchymal stem cells can differentiate into osteoblasts, chondrocytes, myocytes and adipocytes and are the likely source of the amniotic cells in the fibroblast layer which express 11β-HSD1 [23] and also, presumably, of other cells in the amnion, chorion and cytotrophoblast which have also been shown to express the enzyme [24].

Li et al. [25] looked at the relation between cortisol and expression of 11β-HSD1 by IL1-β In cultured human chorionic trophoblast cells. They found that cortisol significantly increased 11β-HSD1 expression and reductase activity. They suggested that the chorion was the most abundant site of 11β-HSD1 within intrauterine tissues. They found that treatment of chorionic trophoblast cells with cortisol (1 μM) induced both cytosolic phospholipase A2 and cyclooxygenase 2 mRNA expression. They concluded that cortisol up-regulated 11β-HSD1 expression through induction of promoter activity, and the effect was enhanced by IL-1, suggesting that more biologically active glucocorticoids could be generated in the fetal membranes in the presence of infection.

5. The metabolism of DHEA

Also located within the endoplasmic reticulum (ER) in many tissues which use 11β-HSD1 to control local glucocorticoid levels is the CYP7B1 enzyme that 7-hydroxylates DHEA. 7OH-DHEA is converted to 7keto-DHEA by 11β-HSD1 [26]. This form of DHEA is more potent for promoting memory function in mice [27]. This 7-hydroxylase enzyme is found in the human placenta with high levels in decidual and trophoblastic cells. The function of this molecule is unknown.

There is a clear difference between the liver hydroxylation of DHEA and that of other tissues. Tissue extracts with the exception of the liver from mice homozygous for targeted disruption of CYP7B fail to generate hydroxylated products of DHEA. The liver has a separate enzyme CYP7A.

This could be very interesting in suggesting that the metabolic needs of the liver are rather different to tissues such as the skin and adipocyte.

Fetal 16-hydroxylation of DHEA and DHEA-S appears to be important [28]. IIt is suggested that steroid sulfates are directly metabolized without prior cleavage of the sulfate group in human fetal livers. The level of DHEA 16α-hydroxylase in human fetal liver is significantly higher than in adult liver. Estriol of fetal origin originates exclusively from the 16- hydroxylation of either DHEA or DHEA-S. The hydroxylated product undergoes aromatization in the syncytiotrophoblast and the estriol then transfers into the maternal circulation [29].

Due to the haemochorial nature of the placenta it is suggested that more than 90% of estriol generated in the syncytiotrophoblast enters the maternal circulation. Nothing is known about a possible interaction between 16OH-DHEA-S and 11β-HSD1 activity. This is important given that cord plasma 16OH-DHEA-S levels are approximately 2.5 times those of DHEA-S.

All this suggests that either the local production and metabolism of DHEA within the ER or its hepatic metabolism may be important. Could it be that metabolites of DHEA and not DHEA itself could affect H6PD expression and hence NADPH production? Thus local conversion of DHEA to 7OH-DHEA and its move to the cytoplasm could produce the non-competitive inhibition of G6PD. This would result in 11β-HSD1 acting as a dehydrogenase and hence converting 7OH-DHEA to 7-keto-DHEA, known to be a more potent neurosteroid and possibly an inhibitor of G6PD.

6. Placental sulfatase

One of the intriguing aspects of this hypothesis is that it potentially explains how it is that DHEA might be biologically relevant even though the circulating levels are two orders of magnitude below those that are necessary to affect NADPH production. It is the generation of DHEA within the ER by the action of steroid sulfatase on DHEA-S that produces very high intracellular DHEA levels. Thus DHEA-S acts as a pro-hormone that is then activated in cells with steroid sulfatase. This enzyme is within the ER and hence ideally placed to generate high levels of DHEA.

Sulfatase activity was first demonstrated in the human placenta by Pulkkinen in 1961 [30] and confirmed by Warren and colleagues [31].

If the hypothesis is correct then placental sulfatase deficiency is of particular interest. Rabe described 76 pregnancies with this [32]. The pregnancies are associated with low oestriol secretion and a prolonged half-life of DHEA-S after venous loading. Only 7 out of 76 pregnancies had an uncomplicated vaginal delivery. Eighteen required Caesarean section. All had a prolonged pregnancy with lack of cervical dilatation and absence of the normal process for the induction of labour. All 76 babies were male. Sulfatase deficiency is a sex-specific, X-linked placental enzyme defect.

7. Why does the placenta have both 11β-HSD1 and 11β-HSD2?

In 1997 we published evidence that cortisol infused into the maternal side of the human placenta was metabolised to cortisone (**Figure 2**) [33]. No conversion of cortisone to cortisol was detected. This study was not designed to examine whether this conversion was solely by 11β-HSD2 or whether 11β-HSD1 could be playing a role. A further more detailed study was published by Sun et al. [34]. They perfused cortisol into the maternal intervillous space of term placentae and found substantial cortisol in the fetal vein. The output of cortisol was increased when carbenoxolone, a non-specific inhibitor of 11β-HSD was added. Of particular interest was their finding that cortisol formation increased in a dose-dependent manner with infusion of cortisone. Cortisol was detected at a low level in both the maternal and fetal outflows. This suggests that *at term* the 11β-HSD1 can act as an oxido-reductase.

In the kidney the inactivation of cortisol to cortisone plays a critically important role in the protection of the mineralocorticoid receptor. The affinity of the glucocorticoid receptor for cortisol is 10–40 times lower than that of the MR. Thus in the kidney there is a need for an extremely effective protective mechanism. Presumably this is also necessary for the placenta in its role of protecting the fetus from maternal cortisol.

Sun et al. looked at the differential expression of 11β-HSD1 and 11β-HSD2 in human placental and fetal membranes [24]. They found that 11β-HSD2 was present in the placenta in the syncytiotrophoblast (**Figure 2**). By contrast 11β-HSD1 was found in the cytotrophoblast, amnion and chorion. 11β-HSD2 was thus well placed to protect the fetus from maternal cortisol. In the mother the circulating levels of free cortisol and cortisone are approximately equal [35].

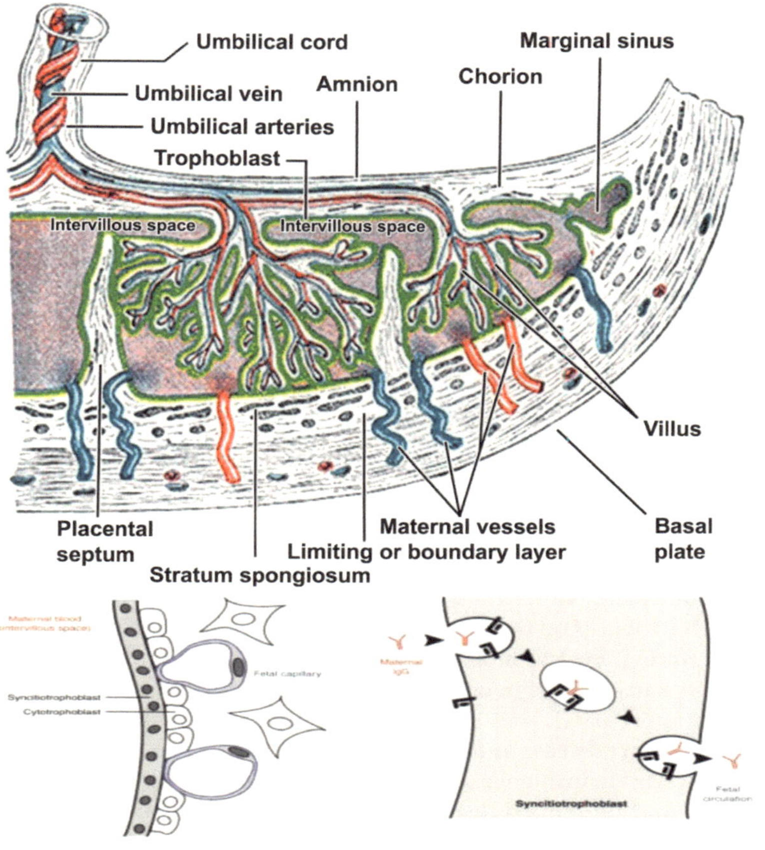

Figure 2.
Diagram of the human placenta. Maternal cortisol enters the inter-villous space and is then metabolised to cortisone by the 11β-HSD2 enzyme in the syncytiotrophoblast. The fetus is either then perfused with cortisone via the umbilical vein or with cortisol if the inactivation is ineffective or the 11β-HSD1 in the amnion, chorion and decidua acts as an oxido-reductase.

Sun et al. [36] suggest that cortisol regeneration in the fetal membranes is not a coincidental but a required event for human parturition. In addition to the role of cortisol in stimulating prostaglandin synthesis (see below) they suggest that local cortisol production induces extracellular matrix (ECM) remodelling and that decreased collagen I, III and IV plays a key role in membrane rupture.

If therefore placental 11β-HSD1 acted normally throughout pregnancy as an oxido-reductase, as most of the literature suggests, it would convert maternal cortisone and cortisone derived from syncytiotrophoblast metabolism to cortisol. This makes little biological sense and is at variance with the studies quoted above which indicate that for the majority of human pregnancy circulating fetal cortisol levels are very low. This suggests that not only is placental 11β-HSD2 effective but that 11β-HSD1 is either down-regulated or acts as a dehydrogenase. This underlines the importance of the mechanism responsible for determining the set point of 11β-HSD1.

Murphy et al. [37] showed that placental 11β-HSD2 activity decreased significantly between 38 and 40 weeks. In contrast there were no changes in 11β-HSD2 mRNA or protein expression.

In comparison 11β-HSD1 mRNA abundance markedly increased with the onset of spontaneous labour as did fetal cortisol concentrations. The presence of the two enzymes in the placenta and 11β-HSD1 in the fetal membranes raises some fascinating questions. Could they function in different ways at different times during pregnancy? Thus placental 11β-HSD2 might be the main way by which the fetus is protected from maternal cortisol but 11β-HSD1 could play an additional local protective role as a dehydrogenase in the majority of the pregnancy (i.e., synergistic with 11β-HSD2). It could then be an oxido-reductase at term increasing *local* (i.e., pericellular and intracellular) cortisol levels at the fetal-maternal interface. This would not necessarily be reflected in measurements of cortisol in the cord blood. This could explain why 11β-HSD1 is located in the cytotrophoblast, chorion and amnion.

Sun and Myatt [23] looked at 11β-HSD1 expression in cultured human amnion cells, the major tissue producing prostaglandins. Term amnion fibroblasts had higher 11β-HSD1 expression than amnion epithelial cells. Dexamethasone dose-dependently increased 11β-HSD1 expression in amnion fibroblasts but not epithelial cells. This was blocked by the GR receptor antagonist RU486. The cytokines IL-1β and TNFα had little effect on 11β-HSD1 expression alone but in combination with dexamethasone significantly enhanced expression. These results suggest that the control of the amniotic glucocorticoid environment could play an important role in the genesis of parturition.

Konstantakou et al. reviewed the 11β-HSD dysregulation during pregnancy and beyond [38]. This focuses on the critical role played by glucocorticoids in developmental programming. Both animal and human studies have confirmed that 11β-HSD2 insufficiency is related to pregnancy complications (pre-eclampsia, intra uterine growth retardation (IUGR), pre-term birth, low birth weight and an increased risk of hypertension, heart disease and type 2 diabetes in adulthood). Konstantakou suggested that the main influence of 11β-HSD1 on fetal physiology is its contribution to fetal lung maturation. This is clearly important but If this was the case it would be difficult to understand why the enzyme is in the placenta and fetal membranes. An interesting study suggested that polymorphisms within the 11β-HSD1 gene conveyed an increased risk of pregnancy-induced hypertension and pre-eclampsia [39].

In summary, 11β-HSD2 with its low Km for cortisol (about 50 nM) (i.e., high affinity) is well placed in the syncytiotrophoblast to convert maternal cortisol to cortisone and thus protect the fetus. By contrast 11β-HSD1 with its higher Km (380 nM for cortisone) is localised to the cytotrophoblast, chorion and amnion and can control local glucocorticoid levels there.

8. Glucocorticoids and prostaglandin production

It might be thought that, given the classical anti-inflammatory effect of glucocorticoids, they would suppress the production of prostaglandins; the reverse appears to be the case. This paradox was noted by Casey et al. [40]. They found that PG production by human amnion and decidual cells in culture was not inhibited by dexamethasone. Indeed the concentration of PGE_2 in the culture medium at day 3 after dexamethasone had been added was doubled. This was very different to the effect of dexamethasone on myometrial smooth muscle cells producing prostacyclin responsible for uterine quiescence. This prostaglandin was markedly inhibited at low dose dexamethasone.

Sun's group delineated the paradoxical mechanism and has admitted that this is very complicated [36]. It involves the glucocorticoid induction of cytosolic phospholipase A2 and COX-2 with subsequent activation of multiple transcription factors.

Others confirmed the paradoxical effect in human amnion and showed that PG stimulation by glucocorticoids induces expression of COX-2 [23]. Amnion fibroblasts produced 50-fold more PGE_2 per cell than epithelial cells. Dexamethasone (0.01–1 μM) increased PGE_2 production in amnion fibroblasts but did not affect PG production in amnion epithelial cells. As indicated above it is the amniotic fibroblasts that contain 11β-HSD1.

Dexamethasone is obviously not an endogenous GR ligand. It is metabolised by 11β-HSD2 to 11-dehydro dexamethasone. 11β-HSD1 can convert this back to dexamethasone but is unable to carry out the reverse dehydrogenase reaction. It is thus important to confirm that cortisol has the same effects. This has been demonstrated by Guo et al. [41]. They also showed that the effect of cortisol on COX-2 in human amnion fibroblasts was modulated by progesterone.

9. Glucocorticoids and oxytocin

Another important possible interaction is between glucocorticoids and oxytocin in amniotic cells. Hinko and Soloff [42] found that oxytocin receptors (OTRs) in rabbit amnion increased 200 fold towards the end of gestation. To study this they first incubated the amnion cells with 20nM cortisol for 24 hours. This increased labelled oxytocin antagonist binding 16-18 fold [43]. *In vivo* experiments confirmed that glucocorticoid treatment induced OTRs in the amnion.

Studies with human amniotic cells suggested that labour and the inflammatory cytokine IL-1β increased the expression of oxytocin receptors [44]. They used primary human amnion cell cultures from patients before and after the onset of labour. The expression of OTR mRNA was 16 fold higher in the post-labour cells than in the pre-labour. Treatment of pre-labour cells with IL-1β brought the level of OTR expression to that observed in post-labour cells. In post-labour cells IL1β produced no significant further increase in OTR.

Previous studies have shown that IL-1β concentrations within the uterus increase at the time of either term or pre-term labour [43]. The same is true for several other so-called labour associated proteins including PTGS2 and IL-8.

9.1 DHEA and nuclear factor kappa B (NF-κB)

Labour, both term and pre-term, is associated with an increase in both the expression and activity of cyclo-oxygenase 2 (PTGS2) in the amnion, the key enzyme involved in the synthesis of prostaglandins [45]. It is regulated by NF-κB [44]. Both basal NF-κB DNA binding and transcriptional activity increase in human amnion at the onset of labour. PTGS2 expression was increased 40 fold and the oxytocin receptor 24 fold. In addition to increased oxytocin receptor mRNA, inflammatory cytokines such as IL-I are found within the amniotic fluid in relation to both term and pre-term labour often related to infection [43]. It is thus of interest that DHEA-S has been shown to markedly inhibit the expression of NF-κB in response to pro-inflammatory cytokines such as TNFα [46]. In the myometrium where abundant Progesterone Receptor B (PRB) is expressed prostaglandin synthesizing enzymes are among the NF-κB mediated pro-inflammatory genes that are inhibited by progesterone during gestation. Is the effect of DHEA-S on NF-κB an additional way in which DHEA can modulate the *in utero* environment?

NF-κB is essential for up-regulation of interleukin-8 expression in human amnion and cervical epithelial cells [47].

9.2 NF-κB and amniotic oxytocin receptors

Terzidou et al. [44] have shown that both labour and inflammation increase the expression of oxytocin receptors (OTRs) in the human amnion. Amniotic prostaglandin synthesis occurs mainly via cyclo-oxygenase 2 that is regulated by NF-κB. Labour, both pre-term and term, is associated with increased expression of this enzyme in the amnion and basal NF-κB DNA binding and transcriptional activity increase with the onset of labour. Using whole genome cDNA arrays they studied the range of genes that have increased expression in amnion that had high levels of NF-κB compared to amnion with low levels [47]. OTR expression showed the second highest difference with 24 fold increase in expression versus 40 fold for COX-2. A total of 919 genes were up-regulated including numerous inflammatory genes. They concluded that amnion activation is largely an inflammatory event that occurs in the amnion epithelial layer as a prelude to the onset of labour.

9.3 Interaction between the NF-κB and C/EBP transcription factor families

Given the very specific effects of pro-inflammatory cytokines such as TNF-α and DHEA on NF-κB and C/EBP transcription factors it is interesting to ask whether they interact. Stein et al. have shown that there is a clear functional and physical interaction [48]. This results in inhibition of promoters with kappa B enhancer motifs and in the synergistic stimulation of promoters with C/EBP binding sites.

10. What controls the production of DHEA-S by the fetal adrenal

If DHEA-S plays regulates placental 11β-HSD1 and 11β-HSD2 then what controls adrenal production of DHEA-S? Mesiano and Jaffe reviewed human fetal adrenal function in pregnancy and parturition [49]. The human placenta produces Corticotrophin Releasing Hormone (CRH) (41 amino acids: Chromosome 8) and the closely related urocortin (40 amino acids: chromosome 2). CRH directly and preferentially stimulates DHEA-S production by fetal adrenal cortical cells [50]. CRH was as effective as ACTH at stimulating DHEA-S production but 70% less potent than ACTH in stimulating cortisol. CRH production was up-regulated by cortisol.

There are two main types of CRH receptor—CRH-R1 and CRH-R2. Sirianni et al. showed that CRH and Urocortin act through type 1 CRH receptors to stimulate DHEA-S production in human fetal adrenal cells [51]. Urocortin mRNA is present in the cytotrophoblast and syncytiotrophoblast [52]. Unlike CRH, urocortin is not secreted into the maternal blood.

Maternal CRH levels rise exponentially as human pregnancy advances [53]. The CRH binding protein (CRH-BP) binds the hormone in an equimolar ratio and prevents the recognition of CRH by the CRH receptor. McLean and colleagues studied this in 485 pregnant women. CRH was detectable in the maternal circulation from early in the second trimester and then rose dramatically. Three hundred and eight women went into spontaneous labour at term: 24 had spontaneous pre-term delivery and 29 were delivered beyond term. The CRH plasma levels were significantly higher in those delivering pre-term and significantly lower in those delivering post-term.

These differences were present from the earliest stage in pregnancy when CRH could be detected. CRH-BP was present greatly in excess of CRH throughout most of pregnancy but the exponential rise in CRH resulted in an equivalence point being reached 20 days before spontaneous term delivery. The CRH-BP levels then decreased significantly so that CRH was present in excess of CRH-BP at serial time points from 20 days before spontaneous delivery. The authors suggest that CRH-BP will block the bioactivity of CRH in the maternal circulation until the final 3 weeks before the onset of spontaneous labour. It is unclear what role CRH-BP might play in the fetal circulation but Perkins et al. have shown that CRH binding in fetal plasma is virtually the same as the maternal samples. If this affects the bioactivity of CRH does this prevent CRH from stimulating the synthesis and release of DHEA-S? Does this suggest that urocortin acting via the adrenal type 1 CRH receptor controls fetal DHEA-S levels?

It is suggested that humans are the only species to have high levels of placental CRH at term and no other species has been reported to have circulating CRH-BP.

Thus if DHEA-S by intra-placental conversion to DHEA controls both the conversion of cortisol to cortisone by 11β-HSD2, as might be the case given the studies of the Odermatt group on this enzyme in the kidney [54] and also the oxido-reductase activity of 11β-HSD1 the following could obtain:

1. Low DHEA-S results in decreased 11β-HSD2 expression and increased 11β-HSD1 conversion of cortisone to cortisol in the placenta.

2. The increase in placental cortisol then enhances the production of placental urocortin that in turn stimulates the fetal adrenal to produce DHEA-S—a classic feedback control system.

Given that the level of CRH rises exponentially during gestation and that the free level of this hormone may be linked to the timing of parturition this poses the question 'What role does placental CRH play in the activation of the fetal hypothalamic pituitary adrenal axis?'. Does free CRH from the maternal circulation cross into the fetus? Is the free level such that it can not only activate the fetal HPA but also directly stimulate cortisol production by the fetal adrenal as shown by Sirianni et al. [51]. If this is so then there are several ways by which circulating cortisol levels in the fetus may rise just before term.

1. Decreased placental 11β-HSD2 activity just before term reduces the fetal protection from maternal cortisol. The transfer increases placental cortisol and hence placental CRH. This then stimulates both the fetal pituitary to secrete ACTH and the adrenal to produce DHEA-S. The ACTH then activates the transitional zone that synthesises cortisol. This further stimulates CRH and hence cortisol.

2. The reduction in maternal CRH-BP results in increased free CRH that crosses the placenta.

11. DHEA-S, cortisol and the onset of labour

Yoon et al. [55] showed that an increase in fetal plasma cortisol but not DHEA-S is followed by the onset of pre-term labour in patients with pre-term premature rupture of the membranes. In those who went into spontaneous labour within 7 days the mean fetal plasma cortisol was 8.35 μg/dl. In contrast those whose onset of labour was beyond 7 days had a mean cortisol of 4.75 µg/dl ($p < 0.0001$). In contrast the DHEA-S levels in the first group had a mean of 154 µg/dl versus 194 µg/dl in the second ($p = 0.09$). This work was confirmed by Oh et al. [56]. They found that fetal plasma cortisol was low until 36 weeks gestation and increased then up to term. Active labour was associated with an increase in plasma cortisol. Fetal plasma DHEA-S increased up to term but not during active labour. What they described as the stress index (cortisol/DHEA-S) increased with progressive gestation and with active labour at term. Others confirmed that infants delivered by Caesarean section had the lowest cortisol levels. Fetal plasma cortisol was significantly elevated in connection with spontaneous human parturition and highest in those born by instrumental delivery [57].

12. The switch from glucocorticoid protection to local glucocorticoid excess

This paper suggests that 11β-HSD enzymes play a key role in protecting the fetus from glucocorticoids. 11β-HSD2 does this by converting maternal cortisol to cortisone in the syncytiotrophoblast and hence together with the failure of cortisol production by the fetal adrenal produces low circulating cortisol levels in the fetus from about 8–10 weeks gestation until just before term. 11β-HSD1 is expressed at low level in the amnion, chorion and decidua and probably acts as a dehydrogenase converting cortisol to cortisone locally within these tissues. DHEA-S produced by the fetal adrenal is then converted to DHEA in the placenta and membranes by placental sulfatase. This local production of DHEA plays a key role in enhancing expression of 11β-HSD2 and decreasing the expression of 11β-HSD1 and H6PDH. Because of the avascular nature of the amnion it is likely that DHEA diffuses through the amnion cells to reach the fibroblasts containing 11β-HSD1.

Key to understanding parturition must very likely involve understanding how the fetus switches from low circulating and low intracellular cortisol levels to higher circulating and very high local glucocorticoid levels which switch on the increased production of oxytocin receptors, prostaglandin E2 and inhibit prostacyclin. The switch does not appear to be related to a decrease in the circulating levels of DHEA-S. This poses the question as how an increase in cortisol or other agents such as pro-inflammatory cytokines might block the inhibitory effect of DHEA.

It is clear that there are a number of key interactions between transcription factors, pro-inflammatory cytokines and glucocorticoids [58]. Grontved looked at the interaction between C/EBP and the Glucocorticoid Receptor (GR). As previously discussed DHEA has a specific effect on the transcription of 11β-HSD1. In particular it decreases the expression of C/EBPα and thus decreases the ratio of C/EBPα/C/EBPβ. What Grontved has shown is that C/EBPβ plays a key role in facilitating glucocorticoid receptor recruitment to DNA steroid response elements. The GR is exclusively associated with accessible chromatin and 62% of GR binding sites are occupied by C/EBPβ. The paper shows that in the liver C/EBPβ frequently occupies the site prior to GR binding and sometimes joins GR at the site after activation. The authors found that accessibility of DNA is highly linked to occupancy of the DNA by C/EBPβ and suggest that C/EBPβ has a priming function for GR binding (**Figure 3**).

Rudiger et al. [59] focus on the fact that the majority of glucocorticoid-modulated genes lack a DNA binding site for the glucocorticoid receptor (GR). They test the hypothesis that besides direct DNA binding of the GR, interaction with other transcription factors modulates the effect of GR binding. Using cultured human mesenchymal cells and leukocytes they found that glucocorticoids induced the formation of a complex

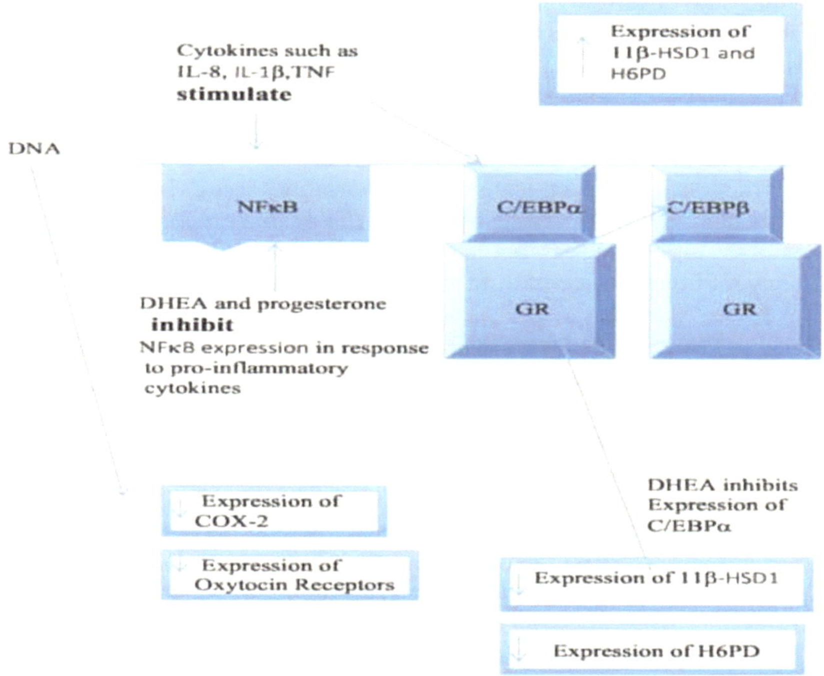

Figure 3.
Interaction of transcription factors with DHEA and pro-inflammatory cytokines with consequent effects on expression of key enzymes and receptors related to parturition.

between the GR and enhancer binding protein C/EBPα. This interaction inhibited glucocorticoid action on mesenchymal and lymphocytic cell proliferation.

13. A comparison of sheep and human parturition

Seminal papers by Liggins and colleagues showed that parturition in sheep is initiated by a sharp rise in the rate of secretion of cortisol by the fetal adrenal [60]. The rise in secretion was thought to be due to a combination of enhanced responsiveness to ACTH and partly to increased fetal ACTH levels. As discussed above cortisol acts in the sheep on placental enzymes active in the biosynthesis of oestrogens from progesterone. Thus placental secretion of oestrogen increases and that of progesterone decreases. They proposed that this change in the ratio of oestrogen: progesterone, particularly the rise in oestrogen, stimulated the release of prostaglandin F2α (PGF2α) from the maternal placenta and myometrium. PGF2α then enhanced the myometrial response to oxytocin and, after a latent period, stimulated contractions. They noted that the onset of parturition is normally associated with softening of the cervix, the mechanism of which is uncertain. The combination of uterine contractions in the presence of a distensible cervix leads to parturition.

In man as already discussed the process has long been thought to be different from that in sheep. In man the placenta was thought to lack the glucocorticoid responsive CYP17 enzyme. This has now been disputed. What this paper suggests is that the fundamental process is cortisol dependent in both species. What differs is the way in which the human fetal adrenal plays a crucial role in controlling the placental and tissue glucocorticoid levels. The effective loss of this control results in a marked increase in the intracellular and peri-cellular levels of cortisol in key tissues secondary to the decreased inactivation of maternal cortisol by placental 11β-HSD2 and the increased expression and oxido-reductase activity of 11β-HSD1 in the amnion, chorion and placenta. This initiates a cascade that stimulates *inter alia* PGE2 production and oxytocin receptor expression. The local release of these hormones results in maturation of the cervix and the onset of myometrial contraction. The same process results in other tissues that express 11β-HSD1 such as the fetal lung increasing local glucocorticoid levels which are key for lung maturation and the production of lung surfactant.

13.1 Pathologies that might affect the modulation of 11β-HSD1 by DHEA

At first sight it might be thought that placental sulfatase deficiency would be a good test of the hypothesis. One might imagine that in such pregnancies the fetus would not be able to produce high concentrations of DHEA within the placenta and thus could not reduce the oxido-reductase activity of 11β-HSD1. The fetus would then be exposed to a much higher *local* concentration of cortisol at sites where 11β-HSD1 is expressed (cytotrophoblast, chorion, amnion). An interesting case study was published in 1976 [61]. A pregnancy with placental sulfatase deficiency presented at 41 weeks and was found to have extremely low urinary oestriol excretion but was otherwise normal. The amniotic fluid DHEA-S concentration was 22-fold higher than in normal pregnancy while that of DHEA was normal. This could be due to the negative feedback control system. Thus failure of stimulation of the expression of 11β-HSD2 activity results in increased intraplacental cortisol and hence increased placental CRH production. This then stimulates DHEA-S and hence the very high levels produced. This can presumably be converted to DHEA by the fetus and could then inhibit placental 11β-HSD1 (i.e., the whole system is now set at a much higher level). At 42 weeks of pregnancy, after induction of labour failed, a healthy normal male infant was delivered by caesarean section. This is in keeping with the 76 male infants described above.

13.2 Pathologies that might affect the modulation of 11β-HSD2

Of particular interest is work showing that cholestatic pregnancy is associated with reduced placental 11β-HSD2 expression [62]. Morris et al. [63] showed that the bile acid chenodeoxycholic acid was a potent inhibitor of 11β-HSD1 dehydrogenase activity. Higher levels were required to inhibit 11β-HSD2. Of special importance was the finding that maternal cholestasis in pregnancy programs metabolic disease in the offspring (males had increased BMI and females increased waist and hip girth) [62]. This was very much in keeping with our studies in pregnant rats given dexamethasone in pregnancy [64]. Given the current global obesity pandemic it would be important to determine what role such neonatal programming is playing in the genesis of the disease.

14. Progesterone

Progesterone plays a key role not only in the maintenance of pregnancy but as a promoter of uterine quiescence. Is there any possibility that it could act in conjunction with DHEA?

In a surprising finding Nadjafi-Triebsch and colleagues discovered that males given DHEA had a rise in their progesterone levels. The mechanism for this was unclear.

Mizutani et al. [65] showed that C/EBPβ plays an important role in progesterone synthesis. These authors found that steroidogenic factor 1 (SF-1) induced the differentiation of mesenchymal stem cells into steroidogenic cells. They then looked at the mechanisms by which SF-1 modulated functions. A key component of the SF-1 nuclear protein complex was C/EBPβ. C/EBPβ is known to be essential for ovulation and luteinisation. C/EBPβ knock down in granulosa cells reduced cAMP- induced progesterone production. They concluded that C/EBPβ is an important mediator of progesterone production by working together with SF-1, especially under trophic hormone-stimulated conditions.

As discussed above DHEA has been shown to down-regulate C/EBPα and hence reduce the C/EBPα/C/EBPβ ratio. These transcription factors act either as homodimers or heterodimers. Does the decrease in C/EBPα lead to an increase in the C/EBPβ homodimer and hence enhance progesterone production as found in the Nadjafi-Triebsch paper?

15. Purinergic receptors in the myometrium

Calcium plays a key role in myometrial contraction and in the oxytocin response [66]. Like in other smooth muscle calcium binds to calmodulin and activates myosin light chain kinase. This can then phosphorylate myosin light chains at serine residue 19. This then allows the myosin crossbridge to bind to actin. The factors that control intracellular calcium in the myometrium are thus of special interest. We demonstrated that aldosterone stimulated the release of ATP from A6 cells [67] which acted on purinergic receptors (P2X4) on the releasing and adjacent cells and thus stimulated calcium entry into the cell. This produced cell contraction and opening of the sodium channel. This could be prevented by adding hexokinase to the cells.

Rat myometrial cells express both P2X4 and P2X7 receptors [68]. The expression of these receptors increased during the late stages of pregnancy. This increase was markedly enhanced in hormone-induced and inflammation-induced pre-term delivery models. Mifepristone increased P2X4 and P2X7 levels 2.1 and 4.1 fold: lipopolysaccharide by 7.4 and 18.6 times the control value.

16. Conclusions

Diczfalusy et al. [69] proposed that the role of the fetal zone of the human fetal adrenal was to produce large amounts of adrenal androgens which were then converted to oestrogens by the placenta. Despite the fact that the role of this oestrogenic milieu is still quite unclear this has been the suggested rationale for the enormous production of DHEA-S (**Figure 4**).

It might seem strange that the FZ develops and DHEA-S secretion starts to increase at the end of the first trimester but the proposed roles of oestrogen are in parturition. One possibility is that cortisol and oestrogens both play a role in parturition that is synergistic (**Figure 5**).

DHEA-S in addition to its function as a source of oestrogens has a vital function as the modulator of 11β-HSD1. Thus for the major part of pregnancy this results in this

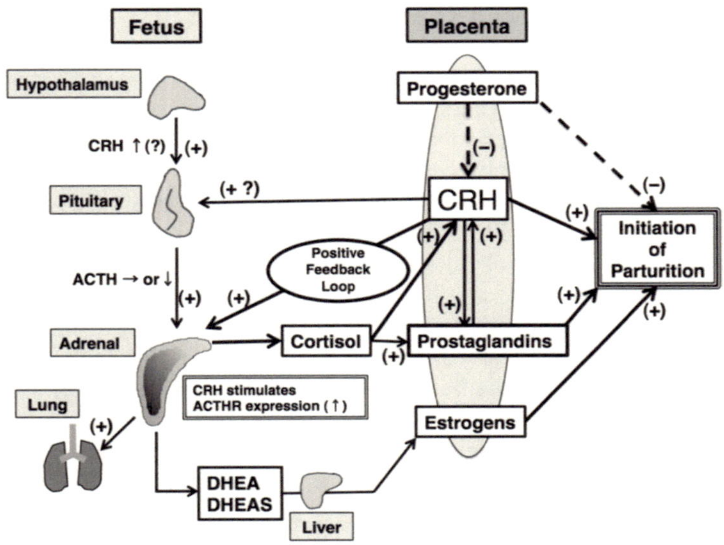

Figure 4.
Classical model of human feto-placental unit (Source: From Ishimoto and Jaffe [3]). Near term, placenta-derived CRH increases and directly stimulates fetal production of DHEA/DHEAS and cortisol. Increased cortisol, in turn, stimulates production of placental CRH. CRH also up-regulates expression of the ACTH Receptor, thereby increasing human fetal adrenal (HFA) responsiveness to ACTH, the level in fetal circulation of which may be constant or decrease near term. HFA and placental CRH constitute a positive feedback cascade. DHEA/DHEAS is converted by the placenta to oestrogen, promoting initiation of parturition partly through up-regulation of contraction-associated proteins. Cortisol also promotes maturation of fetal organs (e.g., the lung) and stimulates production of prostaglandins, uterotonins necessary for parturitional processes. CRH also directly promotes initiation of parturition in part by increasing prostaglandins. When "functional withdrawal" of progesterone occurs, coupled with these changes, parturition is initiated.

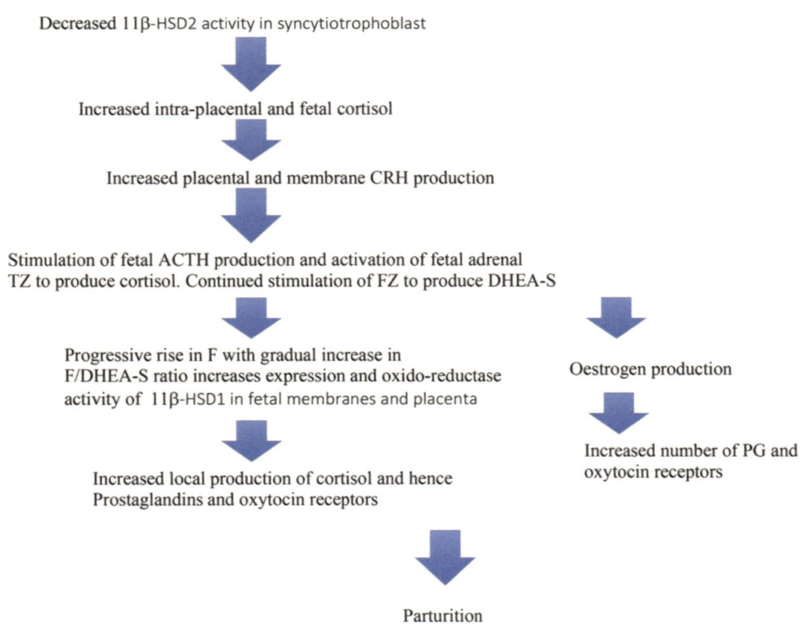

Figure 5.
Possible sequence of events leading to human parturition.

enzyme having both its expression reduced and its function controlled as a dehydrogenase converting cortisol to cortisone. This in conjunction with placental 11β-HSD2 results in a low glucocorticoid environment *in utero* essential for fetal growth. At term this modulation is lost by the effect of the increasing fetal cortisol concentration and results in 11β-HSD1 acting as an oxido-reductase converting cortisone to cortisol in key tissues including the placenta, fetal membranes and the fetal lungs. This then enhances the production of oxytocin receptors and prostaglandins leading to the initiation of labour.

The interaction between DHEA and progesterone leads to a different model of parturition (**Figure 6**).

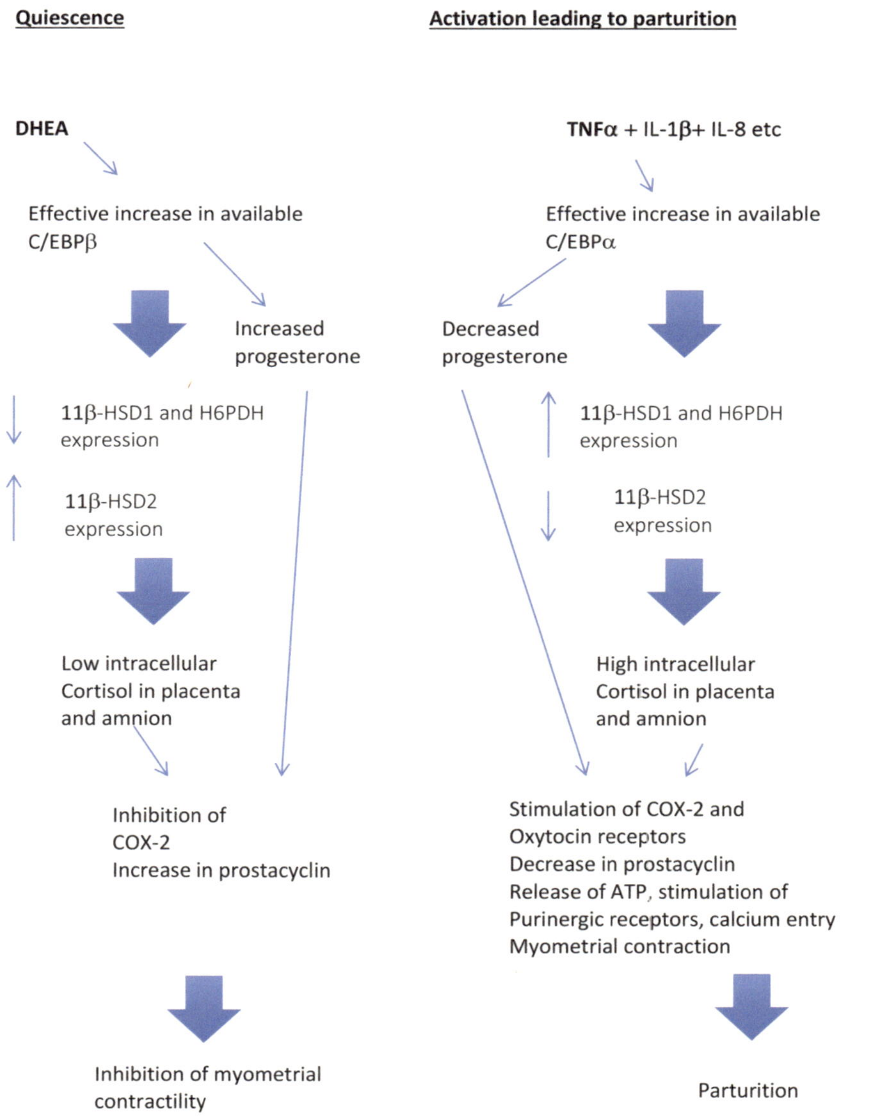

Figure 6.
Model 2 showing the interaction of DHEA with enhancer binding proteins and progesterone producing uterine quiescence and the activation steps leading to parturition.

In the first part of the model DHEA and progesterone are synergistic in producing quiescence during the majority of pregnancy. DHEA increases the C/EBPβ: C/EBPα ratio and thus the availability of C/EBPβ. This together with SF-1 stimulates the synthesis of progesterone which inhibits production of prostaglandins by COX-2. At the same time by the same mechanism DHEA decreases 11β-HSD1 and H6PDH expression resulting in very low amniotic and chorion intracellular cortisol levels. In the myometrium low levels of glucocorticoid enhance the production of prostacyclin and hence uterine quiescence.

In this model parturition is produced, as suggested by Lim et al. [47], by an increase in pro-inflammatory cytokines such as IL-8 TNFα and IL1-β (levels rise in normal pregnancy at term and are significantly higher with infection) that stimulate the production of C/EBPα. This then blocks the inhibitory effect of DHEA on the expression of 11β-HSD1 and H6PDH and also the stimulatory action on 11β-HSD2 expression. As a result there is decreased inactivation of maternal cortisol with increased availability of cortisol within the placenta. This cortisol is synergistic with TNFα and IL1-β in increasing 11β-HSD1 expression and oxido-reductase activity that enables the fetal membranes to convert cortisone to cortisol and thus produce a further major increase of cortisol within the cells. The increased production of C/EBPα induced by inflammatory cytokines inhibits prostaglandin synthesis which is dependent on C/EBPβ. This combination of an increase in local cortisol concentrations coupled with the decrease in progesterone within the cell results in COX-2 induction and hence prostaglandin production. The local glucocorticoid production also stimulates a marked increase in oxytocin receptors, inhibits uterine prostacyclin production and hence enhances myometrial contractility. Myometrial contraction is further enhanced by TNFα and IL1-β increasing the expression of P2X4 and P2X7 receptors in the myometrium. These may play an important role in the elevation of intracellular calcium induced by oxytocin.

Author details

Christopher Edwards
Department of Medicine, Imperial College, London, UK

*Address all correspondence to: c.edwards@ncl.ac.uk

References

[1] Escobar JC, Patel SS, Beshay VE, Suzuki T, Carr BR. The human placenta expresses CYP17 and generates androgens *De Novo*. The Journal of Clinical Endocrinology and Metabolism. 2011;**96**(5):1385-1392

[2] Hong SH et al. Expression of steroidogenic enzymes in human placenta according to the gestational age. Molecular Medicine Reports. 2019;**49**(5):3903-3911

[3] Ishimoto H, Jaffe RB. Development and function of the human fetal adrenal cortex: A key component in the feto-placental unit. Endocrine Reviews. 2011;**32**(3):317-355

[4] Pignatti E, du Toit T, Flück CE. Development and function of the fetal adrenal. Reviews in Endocrine & Metabolic Disorders. 2023;**24**(1):5-21

[5] Goto M et al. In humans, early cortisol biosynthesis provides a mechanism to safeguard female sexual development. The Journal of Clinical Investigation. 2006;**116**(4):953-960

[6] Murphy BEP. Human fetal serum cortisol levels related to gestational age: Evidence of a midgestational fall and a steep late gestational rise, independent of sex or mode of delivery. American Journal of Obstetrics and Gynecology. 1982;**144**(3):276-282

[7] Edwards CR et al. Localisation of 11 beta-hydroxysteroid dehydrogenase—Tissue specific protector of the mineralocorticoid receptor. Lancet. 1988;**2**(8618):986-989

[8] Stewart PM, Corrie JE, Shackleton CH, Edwards CR. Syndrome of apparent mineralocorticoid excess. A defect in the cortisol-cortisone shuttle. The Journal of Clinical Investigation. 1988;**82**(1):340-349

[9] Bujalska IJ et al. Hexose-6-phosphate dehydrogenase confers oxo-reductase activity upon 11 beta-hydroxysteroid dehydrogenase type 1. Journal of Molecular Endocrinology. 2005;**34**(3):675-684

[10] Barash V, Erlich T, Bashan N. Microsomal hexose-6-phosphate and 6-phosphogluconate dehydrogenases in extrahepatic tissues: Human placenta and pig kidney cortex. Biochemistry International. 1990;**20**(2):267-274

[11] Marks PA, Banks J. Inhibition of mammalian glucose-6-phosphate dehydrogenase by steroids. Proceedings of the National Academy of Sciences of the United States of America. 1960;**46**(4):447-452

[12] Raineri R, Levy HR. On the specificity of steroid interaction with mammary glucose 6-phosphate dehydrogenase. Biochemistry. 1970;**9**(11):2233-2243

[13] Ziboh VA, Dreize MA, Hsia SL. Inhibition of lipid synthesis and glucose-6-phosphate dehydrogenase in rat skin by dehydroepiandrosterone. Journal of Lipid Research. 1970;**11**(4):346-354

[14] McCormick KL, Wang X, Mick GJ. Evidence that the 11 beta-hydroxysteroid dehydrogenase (11 beta-HSD1) is regulated by pentose pathway flux. Studies in rat adipocytes and microsomes. The Journal of Biological Chemistry. 2006;**281**(1):341-347

[15] Haschemi A et al. Article the sedoheptulose kinase CARKL directs

macrophage polarization through control of glucose metabolism. Cell Metabolism. 2012;**15**(6):813-826

[16] Senesi S, Legeza B, Balazs Z, Csala M, Odermatt A. Contribution of fructose-6-phosphate to glucocorticoid activation in the endoplasmic reticulum: Possible implication in the metabolic syndrome. Endocrinology. 2010;**151**(October): 4830-4839

[17] McNelis JC et al. Dehydroepiandrosterone exerts antiglucocorticoid action on human preadipocyte proliferation, differentiation, and glucose uptake. American Journal of Physiology. Endocrinology and Metabolism. 2013;**305**(9):E1134-E1144

[18] Apostolova G, et al. Dehydroepiandrosterone Inhibits the Amplification of Glucocorticoid Action in Adipose Tissue. American Journal of Physiology-Endocrinology and Metabolism. 2005;**288**:E957-E964

[19] Balazs Z, Schweizer RAS, Frey FJ, Rohner-Jeanrenaud F, Odermatt A. DHEA induces 11 -HSD2 by acting on CCAAT/enhancer-binding proteins. Journal of American Society of Nephrology. 2008;**19**(1):92-101

[20] Chapman KE, Coutinho AE, Gray M, Gilmour JS, Savill JS, Seckl JR. The role and regulation of 11β-hydroxysteroid dehydrogenase type 1 in the inflammatory response. Molecular and Cellular Endocrinology. 2009;**301**(1-2):123-131

[21] Hardy RS et al. Differential expression, function and response to inflammatory stimuli of 11beta-hydroxysteroid dehydrogenase type 1 in human fibroblasts: A mechanism for tissue-specific regulation of inflammation. Arthritis Research & Therapy. 2006;**8**(4):R108

[22] Ahasan MM et al. Inflammatory regulation of glucocorticoid metabolism in mesenchymal stromal cells. Arthritis and Rheumatism. 2012;**64**(7):2404-2413

[23] Sun K, Myatt L. Enhancement of glucocorticoid-induced 11β-hydroxysteroid dehydrogenase type 1 expression by proinflammatory cytokines in cultured human amnion fibroblasts. Endocrinology. 2003;**144**(12):5568-5577

[24] Sun K, Yang K, Challis JRG. Differential expression of 11β-hydroxysteroid dehydrogenase types 1 and 2 in human placenta and fetal membranes 1. The Journal of Clinical Endocrinology and Metabolism. 1997;**82**(1):300-305

[25] Li W, Gao L, Wang Y, Duan T, Myatt L, Sun K. Enhancement of cortisol-induced 11β-hydroxysteroid dehydrogenase type 1 expression by interleukin 1β in cultured human chorionic trophoblast cells. 2006;**147**(5):2490-2495

[26] Nashev LG et al. Hexose-6-phosphate dehydrogenase modulates 11beta-hydroxysteroid dehydrogenase type 1-dependent metabolism of 7-keto- and 7beta-hydroxy-neurosteroids. PLoS One. 2007;**2**(6):e561

[27] Yau JLW et al. Dehydroepi-androsterone 7- hydroxylase CYP7B: Predominant expression in primate hippocampus and reduced expression in Alzheimer's disease. Neuroscience. 2003;**121**(2):307-314

[28] Kitada M, Kamatakis T, Itahashiq K, Rikihisa T, Kanakubo Y. P-450 HFLa, a form of cytochrome P-450 purified from human fetal livers, is the 16~hydroxylase of dehydroepiandrosterone 3=Sulfate*. The Journal of Biological Chemistry. 1987;**262**:13534-13537

[29] Falah N, Torday J, Quinney SK, Haas DM. Estriol review: Clinical applications and potential biomedical importance. Clinical Research Trials. 2015;**1**(2):29-33

[30] Pulkkinen M. Arylsulphatase and the hydrolysis of some steroid sulphates in developing organism and placenta. Acta Physiologica Scandinavica. Supplementum. 1961;**52**(180):1-92

[31] French AP, Warren JC. Sulfatase activity in the human placenta. Steroids. 1966;**8**(1):79-85

[32] Rabe T, Hösch R, Runnebaum B. Sulfatase deficiency in the human placenta: Clinical findings. Biological Research in Pregnancy and Perinatology. 1983;**4**(3):95-102

[33] Benediktsson R, Calder AA, Edwards CRW, Seckl JR. Placental 11β-hydroxysteroid dehydrogenase: A key regulator of fetal glucocorticoid exposure. Clinical Endocrinology. 1997;**46**(2):161-166

[34] Sun K, Adamson SL, Yang K, Challis JR. Interconversion of cortisol and cortisone by 11beta-hydroxysteroid dehydrogenases type 1 and 2 in the perfused human placenta. Placenta. 1999;**20**(1):13-19

[35] Walker BR, Connacher AA, Lindsay RM, Webb DJ, Edwards CR. Carbenoxolone increases hepatic insulin sensitivity in man: A novel role for 11-oxosteroid reductase in enhancing glucocorticoid receptor activation. The Journal of Clinical Endocrinology and Metabolism. 1995;**80**(11):3155-3159

[36] Wang W, Guo C, Sun K. Cortisol regeneration in the fetal membranes: A coincidental or requisite event in human parturition? Frontiers in Physiology. 2020;**11**(May):1-10

[37] Murphy VE et al. Alterations in human placental 11beta-hydroxysteroid dehydrogenase type 1 and 2 with gestational age and labour. Placenta. 2003;**24**(7):739-744

[38] Konstantakou P, Mastorakos G, Vrachnis N, Tomlinson JW, Valsamakis G. Dysregulation of 11beta-hydroxysteroid dehydrogenases: Implications during pregnancy and beyond. Journal of Maternal and Neonatal Medicine. 2017;**30**(3):284-293

[39] Shimodaira M et al. Glucocorticoid synthesis-related genes: *HSD11B1* and *HSD11B2* in hypertensive disorders in pregnancy. Gynecological Endocrinology. 2013;**29**(7):657-661

[40] Casey ML, MacDonald PC, Mitchell MD. Despite a massive increase in cortisol secretion in women during parturition, there is an equally massive increase in prostaglandin synthesis. A paradox? The Journal of Clinical Investigation. 1985;**75**(6):1852-1857

[41] Guo CM, Zhu XO, Ni XT, Yang Z, Myatt L, Sun K. Expression of progesterone receptor a form and its role in the interaction of progesterone with cortisol on cyclooxygenase-2 expression in amnionic fibroblasts. The Journal of Clinical Endocrinology and Metabolism. 2009;**94**(12):5085-5092

[42] Hinko A, Soloff MS. Up-regulation of oxytocin receptors in rabbit amnion by glucocorticoids: Potentiation by cyclic adenosine 3′,5′-monophosphate. Endocrinology. 1993;**133**(4):1511-1519

[43] Romero R et al. Interleukin-1 alpha and interleukin-1 beta in preterm and term human parturition. American Journal of Reproductive Immunology. **27**(3-4):117-123

[44] Terzidou V, Blanks AM, Kim SH, Thornton S, Bennett PR. Labor and

inflammation increase the expression of oxytocin receptor in human amnion. Biology of Reproduction. 2011;**84**(3):546-552

[45] Zakar T, Olson DM, Teixeira FJ, Hirst JJ. Regulation of prostaglandin endoperoxide H2 synthase in term human gestational tissues. Acta Physiologica Hungarica. 1996;**84**(2):109-118

[46] Altman R, Motton DD, Kota RS, Rutledge JC. Inhibition of vascular inflammation by dehydroepiandrosterone sulfate in human aortic endothelial cells: Roles of PPARα and NF-κB. Vascular Pharmacology. 2008;**48**(2-3):76-84

[47] Lim S et al. Nuclear factor kappa B activation occurs in the amnion prior to labour onset and modulates the expression of numerous labour associated genes. PLoS One. 2012;**7**(4):e34707

[48] Stein B, Cogswell PC, Baldwin AS. Functional and physical associations between NF-kappa B and C/EBP family members: A Rel domain-bZIP interaction. Molecular and Cellular Biology. 1993;**13**(7):3964-3974

[49] Mesiano S, Jaffe RB. Developmental and functional biology of the primate Fetal adrenal cortex [1]. Endocrine Reviews. 1997;**18**(3):378-403

[50] Smith R, Mesiano S, Chan E-C, Brown S, Jaffe RB. Corticotropin-releasing hormone directly and preferentially stimulates dehydroepiandrosterone sulfate secretion by human fetal adrenal cortical cells. The Journal of Clinical Endocrinology and Metabolism. 1998;**83**(8):2916-2920

[51] Sirianni R, Mayhew BA, Carr BR, Parker CR, Rainey WE. Corticotropin-releasing hormone (CRH) and urocortin act through type 1 CRH receptors to stimulate dehydroepiandrosterone sulfate production in human fetal adrenal cells. The Journal of Clinical Endocrinology and Metabolism. 2005;**90**(9):5393-5400

[52] Watanabe F et al. Urocortin in human placenta and maternal plasma. Peptides. 1999;**20**(2):205-209

[53] McLean M, Smith R. Corticotropin-releasing hormone in human pregnancy and parturition. Trends in Endocrinology and Metabolism. 1999;**10**(5):174-178

[54] Balazs Z, Schweizer RAS, Frey FJ, Oise Rohner-Jeanrenaud F, Odermatt A. DHEA induces 11β-HSD2 by acting on CCAAT/enhancer-binding proteins. Journal of American Society Nephrology. 2008;**19**:92-101

[55] Yoon BH et al. An increase in fetal plasma cortisol but not dehydroepiandrosterone sulfate is followed by the onset of preterm labor in patients with preterm premature rupture of the membranes. American Journal of Obstetrics and Gynecology. 1998;**179**(5):1107-1114

[56] Oh S-Y, Romero R, Shim S-S, Park JS, Jun JK, Yoon BH. Fetal plasma cortisol and dehydroepiandrosterone sulfate concentrations in pregnancy and term parturition. Journal of Maternal and Neonatal Medicine. 2006;**19**(9):529-536

[57] Mears K, McAuliffe F, Grimes H, Morrison J. Fetal cortisol in relation to labour, intrapartum events and mode of delivery. Journal of Obstetetrics and Gynaecology (Lahore). 2004;**24**(2):129-132

[58] Grøntved L et al. C/EBP maintains chromatin accessibility in liver and facilitates glucocorticoid receptor recruitment to steroid response elements

transaction report. EMBO Journal. 2013;**32**:1568-1583

[59] Rudiger JJ. Interaction of C/EBPalpha and the glucocorticoid receptor in vivo and in nontransformed human cells. The FASEB Journal. 2002;**16**(2):177-184

[60] Liggins GC, Fairclough RJ, Grieves SA, Forster CS, Knox BS. Parturition in the sheep. Ciba Foundation Symposium. 1977;**47**:5-30

[61] Osathanondh R, Canick J, Ryan KJ, Tulchinsky D. Placental sulfatase deficiency: A case study. The Journal of Clinical Endocrinology and Metabolism. 1976;**43**(1):208-214

[62] Martineau M et al. Cholestatic pregnancy is associated with reduced placental 11βHSD2 expression. Placenta. 2014;**35**(1):37-43

[63] Morris DJ, Souness GW, Latif SA, Hardy MP, Brem AS. Effect of chenodeoxycholic acid on 11β-hydroxysteroid dehydrogenase in various target tissues. Metabolism. 2004;**53**(6):811-816

[64] Benediktsson R, Lindsay RS, Noble J, Seckl JR, Edwards CRW. Glucocorticoid exposure in utero: New model for adult hypertension. The Lancet. 1993;**341**(8841):339-341

[65] Mizutani T et al. C/EBPβ (CCAAT/enhancer-binding protein β) mediates progesterone production through transcriptional regulation in co-operation with SF-1 (steroidogenic factor-1). The Biochemical Journal. 2014;**460**(3)

[66] Arrowsmith S, Wray S. Oxytocin: Its mechanism of action and receptor signalling in the myometrium. Journal of Neuroendocrinology. 2014;**26**(6):356-369

[67] Gorelik J et al. Aldosterone acts via an ATP autocrine/paracrine system: The Edelman ATP hypothesis revisited. Proceedings of the National Academy of Sciences of the United States of America. 2005;**102**(42):15000-15005

[68] Urabe S, Miyoshi H, Fujiwara H, Yamaoka K, Kudo Y. Enhanced expression of P2X4 and P2X7 purinergic receptors in the myometrium of pregnant rats in preterm delivery models. Reproductive Sciences. 2009;**16**(12):1186-1192

[69] Diczfalusy E. A modified theory of steroid synthesis in the human foeto-placental unit. In: Endokrinologie der Entwicklung und Reifung. Berlin, Heidelberg: Springer; 1970. pp. 32-46